The Medical Science

of

Total Body Transformation

Neeraj Goswamy, MD

ISBN: 978-1-4834-1826-1 (sc)
ISBN: 978-1-4834-1825-4 (e)

Lulu Publishing Services rev. date: 09/22/2014

In memory of my grandmother

Kaushalya Devi Goswamy
October 18, 1920–October 25, 1985
without whom this book could never have been created

Contents

Chapter 1

My Total Body Transformation

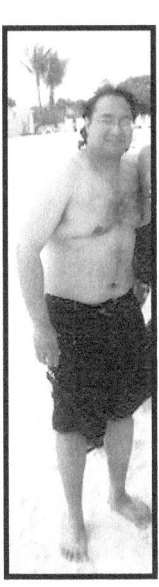

Pre Total Body Transformation
The subject (me)
Weight: 200 lbs.
Waistline: 36"
Abdominal circumference at the level of the umbilicus: 40"
Body fat percentage: 27%
BMI: 30.9
With hypertension, hyperlipidemia, obesity, snoring, and backache

I was in good health in my childhood and was very thin for my age. At this stage in my life I was eating more and more home cooked Indian food. This food is less processed and mostly vegetarian. I was raised by my grandmother, who never touched meat.

When I started eating foods that were not home cooked, I found myself getting fatter, weaker, and more out of shape. I tried a variety of exercises and activities, such as running, rollerblading, sit-ups, push-ups, and so forth. However, I still did not see any results. At the age of 29, I purchased my Bowflex and tried to take working out more seriously. I chose the Bowflex because it combined strength training with an aerobic workout. There was no momentum with the resistance, so I could work out as fast as I could. Weights are bad for your joints, but Bowflex is not. At age 29, I was 5 feet 7½ inches tall and weighed about 180 pounds. I started out bench pressing, chest flying, and shoulder pressing with 80 pounds of resistance; I was also arm curling 80 pounds of resistance. My workout would last about an hour.

By age 32, I was able to bench press, chest fly, and shoulder press 410 pounds of resistance, the maximum; I could also arm curl 310 pounds. I was able to continue doing this until about age 37. By then, I weighed 200 pounds and had put on a significant amount of muscle. After all, you can't go from bench-pressing 80 to 410 pounds without putting on muscle. However, I was unable to lose the fat. My waistline remained at 36 inches, and my abdominal circumference at the level of the umbilicus remained at 40 inches. My body fat percentage was 27, which classified me as obese. I had a history of heart disease in my family; my grandfather and father had both had heart attacks at the age of 40. An abdominal circumference of 40 inches would put anyone at the risk of a heart attack. Due to my abdominal circumference and family history, I was definitely headed for a heart attack by the age of 40. Any exercise I tried at this point didn't help. Nothing I did helped me lose weight.

So, at 37, I weighed 200 pounds, had a body fat measurement of 27 percent, and my abdominal circumference at the level of the umbilicus was 40 inches. At this point in my life, I started focusing more on my diet, because it was my last resort. I started to read labels more and was pickier about the food I ate. I completely changed my diet. I found myself going back to my textbooks from medical school. I was successful in losing 81 pounds in 7½ months and went down to a weight of 119 pounds. My body fat percentage was around 4–5 percent. My waistline went down to 29 inches, and my abdominal circumference at the level of the belly button was 27 inches. I had lost 13 inches around my umbilicus and 8 inches around my waist. Seven and a half months of attention to my diet did more for me than a lifetime of exercise. I went from 200 pounds to 119 pounds. My hypertension and snoring went away. So did my backache, my obesity, and my hyperlipidemia. I am pretty sure I also prevented a heart attack, stroke, diabetes, and a host of other problems. I did only light aerobic exercise for 20 minutes twice a day, once in the morning and once at night before going to bed. When I was running three miles a day, my weight went from 150 pounds to 130 pounds. Then I stopped running because I hurt my knee. So it is quite obvious that most of my weight loss was attributed to diet and not exercise. My goals were extreme. I wanted to go from being obese to having six-pack abs. I learned that the six-pack abs wouldn't be achieved in the gym, but in the kitchen.

My total body transformation consisted of a *catabolic phase*, when I burned off fat; this lasted 7½ months. Then there was an *anabolic phase*, when I built up muscle and was on a high-protein diet. This lasted two weeks. I put on 13 pounds of muscle. The most important thing in this transformation was not strength or endurance, but strict discipline. One can get away with not working out but eating the right diet and still lose weight, but not vice versa. It is very important to eat properly when it comes to losing weight. Exercise can speed up the weight loss, but it is not really necessary. Without exercise, the weight loss will go at a slower pace. Whether you are trying to lose fat or inches, the most important factor is your diet, not exercise. The most important point that I will make here and throughout this book is that *exercise cannot be a substitute for a good diet.*

The way to lose body fat is by depleting the body of its fat stores. This is done by stopping the intake of fats and anything that can be converted and stored as body fat. This will take off body fat faster than any exercise. The typical answer to the question "How do I lose weight?" is *diet and exercise.* It is easier to exercise than to change your diet, because changing your diet requires a lifestyle change. So most people try to lose weight just by exercising, and when it doesn't work they stop exercising. Other people go on temporary diets like Atkins, which have no real permanent effects. Building up muscle requires strength training and a high protein diet. Again, it may take longer for some people. It took me two weeks to regain the strength I had before my total body transformation. I had been strength training for about eight years before my total body transformation. This is why my anabolic phase lasted only two weeks. Another reason is that after I brought down my body fat percentage to 4 percent, my testosterone level increased. This helped me build up muscle mass quickly. I strength trained twice a day for those two weeks.

The biggest benefit of my total body transformation is seen in my lipid profile. I used to have hyperlipidemia. After my transformation, my LDL, triglycerides, and cholesterol all went to the lower end of the normal range. My HDL was above normal range. A favorable lipid profile can prevent a lot of health problems.

It is one thing to write a book about total body transformation when you have been in shape all of your life. It is another thing to write it from the point of view of someone who was unhealthy and out shape. I decided to make a change. If you are unhealthy and out of shape and you are ready to make a change, then this book is definitely for you. I am a physician as a well a certified master fitness trainer. I used established medical science, my own personal experience, and independent research to put this book together. I know the pitfalls and plateaus of a total body transformation because I have been there. I can guide you down the same path that I went down if you are ready.

Always remember to check with your physician before you participate in any weight loss program, muscle-building workout, diet, supplement program, or any other regimen.

Chapter 2

Basic Nutrition and Biochemistry

A s we know, the major nutrients are fats, carbohydrates, and proteins. The common myth in the '80s and '90s was that we needed to stay away from fats. Then, from 2000 to about 2010, people thought we needed to stay away from carbohydrates. The fact is that the body needs all three classes of nutrients—fats, carbohydrates, and proteins— in order to lose weight and remain healthy.

We will first talk about fats. Twenty to 35 percent of total calories should come from fats. Fats have nine calories per gram, and there are three different types: saturated fats, trans fats, and unsaturated fats. Some of these fats break down body fat, while others increase body fat. Body fat is stored in the *adipocytes* or fat cells of the body, and also in the muscle.

Saturated fats are solid at room temperature. They are solid because their chemical structure does not contain any double bonds. They are found in animal products, such as beef, pork, bacon, ham, lard, whole milk, whole cheese, sour cream, and so forth. They are also found in coconut, butter, coconut oil, and palm oil. These fats are known to increase your bad cholesterol or LDL. Some say they also increase your HDL as well. Saturated fats increase your body fat.

Then there are three types of unsaturated fats: *trans fats*, *monounsaturated fats*, and *polyunsaturated fats*. Trans fats are unsaturated fats that are partially hydrogenated. When they become completely hydrogenated they lose their double bond and become saturated fats. Trans fats typically have carbon atoms on opposite sides of the double bond, which makes them have a straight structure. *Cis* fats have carbon atoms on the same side of the double bond, which gives them a crooked structure. The monounsaturated fats discussed in the next paragraph are *cis* fats. Although trans fats have a double bond, they behave like saturated fats and are solid at room temperature. Trans fats are the worst kind of fats. They increase your bad cholesterol (LDL) and decrease your good cholesterol or (HDL). Like saturated fats, trans fats are known to increase your body fat. Many foods contain trans fats, including: frozen foods, fast foods, margarine, cake mixes, soups, fried foods (such as french fries and fried chicken), certain pies, doughnuts, cookies, chips, crackers, cereals, candies, toppings, cake icing, dips, salad dressings, and others. They are found more in processed foods. Two naturally occurring trans fats are *vaccenic acid* and *conjugated linoleic acid* (CLA). They are found in the fat and milk of ruminants. CLA is actually a supplement used for weight loss and will be discussed later in this book.

Two other unsaturated fats are *monounsaturated fats* and *polyunsaturated fats*. These fats are liquids at room temperature. They are liquid because they have one or more double bonds. Monounsaturated fats have one double bond in *cis* configuration. They increase good cholesterol (HDL) and decrease bad cholesterol (LDL). Some sources say they may leave HDL unchanged. This would still be a favorable effect because it lowers LDL. Monounsaturated fats are found in milk, peanuts, cashews, avocados, safflower oil, olive oil, canola oil, almonds, pecans, pumpkinseeds, sunflower seeds, and so forth. Polyunsaturated fats have more than one double bond, and they decrease your bad cholesterol (LDL). Other sources say they have no effect on LDL but that they raise your HDL. Others believe they lower both your HDL and LDL. Omega 3 fatty acids (found in fish) are in this category, and they include the essential fatty acid *linolenic acid*. Foods that contain polyunsaturated fats are fish oil, walnuts, salmon, tuna, sardines, and others. Many of the foods that have monounsaturated fats also have polyunsaturated fats. Both of these fats are known to decrease body fat.

We will now talk about carbohydrates. The recommended dietary allowance for carbohydrates is 130 grams a day. Forty-five to 65 percent of the calories should be coming from carbohydrates. Less than 25 percent of the calories should come from added sugars. Carbohydrates have four calories per gram and can be classified as either *simple* or *complex*. Simple carbohydrates are simple sugars, either *monosaccharides* or *disaccharides*. Examples of monosaccharides—the sugars our bodies can absorb—are glucose, fructose, and galactose. Monosaccharides originate from disaccharides, which are made up of two sugar molecules. Examples of disaccharides are sucrose (table sugar), maltose, and lactose. Sucrose is made up of a molecule of glucose and fructose, lactose is made up of molecules of glucose and galactose, and maltose is made by joining two glucose molecules together. Simple carbohydrates are quickly broken down by the body and are rapidly absorbed into the bloodstream. They create a high glucose level, which results in an insulin spike. Because they create this high glucose level, they are known to have a high glycemic index. When insulin is released, it causes the sugar to be taken up into the liver and muscles. Sugar is stored in the liver and the muscles in the form of glycogen. Once the liver and muscles have exceeded their capacity to store sugar, the sugar then becomes converted to fat and is stored as body fat. The body can store unlimited quantities of fat; it does have the capability of synthesizing sugar from protein via a process called *gluconeogenesis*. However, the body cannot convert stored fat back into sugar. Plants can do this but not humans. Insulin is also known to prevent the breakdown of body fat. So, simple carbohydrates not only prevent the breakdown of body fat, but also they increase body fat. Examples of foods with simple carbohydrates are cakes, cookies, candies, fruit juices, table sugar, honey, soft drinks, cakes, syrup, jam, chocolate, and so forth. Essentially, anything that tastes sweet has simple carbohydrates.

Let's talk about complex carbohydrates. Complex carbohydrates are known as *polysaccharides*. They are composed of numerous long chains of sugar molecules. For nutritional purposes, there are two types of polysaccharides, starch and fiber. Complex carbohydrates do not necessarily taste sweet. Starch is composed of numerous long chains of glucose molecules. Starch is soluble in water and is broken down into glucose by the body. Some foods that contain starch and are low in fiber are potatoes, rice, white bread, cereals, and pasta.

Then there is fiber, another complex carbohydrate. Men should consume 38 grams of fiber and day, and women should consume 25 grams. The typical American diet has only 11 grams of fiber in it. Fiber is a polysaccharide that is non-starch. Examples of fiber are lignin, inulin, cellulose, arabinoxylans, waxes, chitins, pectins, oligosaccharides, and beta-glucans. Oligosaccharides are composed of two to ten molecules of sugar. They can result from the digestion of polysaccharides. Fiber is often broken down incompletely by the body and is not very digestible. There are two kinds of fiber: soluble and insoluble. Soluble fiber affects the way other nutrients are absorbed. Fiber lowers cholesterol, prevents fat absorption, and also decreases the glycemic response. Insoluble fiber usually just moves through the GI tract and absorbs water. It acts as a bulking agent for the stool. Some foods that contain fiber are fruits, vegetables, nuts, beans, lentils, and grains.

High fiber foods tend to have a low glycemic index. This is because it takes the body time to break them down, so these foods tend to release glucose into the bloodstream slowly. This increases the blood sugar slightly, but will not cause a blood sugar spike—so insulin is released in lesser amounts. The blood needs to have a certain level of sugar in it in order for the brain and red blood cells to function. When we don't eat, a constant supply of sugar is released into the blood by glycogenolysis and gluconeogenesis, processes that occur in the liver. Our liver produces the sugar needed for brain and red blood cells to function between meals. This sugar is also used up by the body for various things by various organs. When the sugar runs out, the brain uses ketone bodies. The brain can only live off glucose and ketone bodies; it cannot live off fats or proteins. Ketone bodies form when the body does not have glucose. A drop in blood sugar can cause someone to faint. The hormones that break down fat and sugar are glucagon, growth hormone, and adrenaline. The hormone that increases stored fat and causes sugar to enter the cells is insulin. At a blood sugar level of 85 mg/dl, insulin production decreases. Glucagon, adrenaline, and growth hormone production begins at a blood sugar level of 65–70 mg/dl. Hypoglycemic symptoms begin at a blood sugar of 55 mg/dl. So it is not a good idea to eat foods with a high glycemic index, because they will cause the release of insulin. You want to have enough sugar in the bloodstream to avoid hypoglycemic symptoms and get the beneficial fat-burning effect of glucagon, growth hormone, and adrenaline—but you do not want to have too much, which would cause the release of insulin. So, starchy foods that are low in fiber (such as the ones listed above) have a high glycemic index, and foods that are high in fiber have a low glycemic index. Foods that have a high glycemic index generally prevent fat breakdown and increase stored body fat, whereas foods that have a low glycemic index do just the opposite. A low glycemic index would be below fifty-five. A medium glycemic index would be between fifty-six and sixty-nine. A high glycemic index would be above seventy. So some fruits and vegetables may be high in simple sugars, but because they have fiber in them, their glycemic index is low and eating them will not result in an increase in body fat.

Some food labels may contain a type of carbohydrate called sugar alcohols. Sugar alcohols are neither sugars nor alcohols; they are substitutes for sugars. Many sugar free foods contain sugar alcohols, as they are used as sweeteners. Some examples of sugar alcohols are maltitol, xylitol, erythritol, sorbitol, mannitol, arabitol, and lactitol. Sugar alcohols are incompletely broken down in the gut and are not fully absorbed. They have a low glycemic index and do not cause an insulin spike. They can cause gas, bloating, and diarrhea. While some sources report that they do not cause weight gain, some people have reported weight gain while consuming them. They are often part of diabetic food items.

The third major nutrient is protein. Protein has four calories per gram. Proteins are made up of 20 amino acids. Amino acids are classified as essential and nonessential (Table 1). Essential acids are amino acids that are not naturally synthesized by the body and have to be obtained from an outside source (such as meats, dairy, fruits, or vegetables), and nonessential amino acids are synthesized by the body from metabolic precursors. Protein is necessary to help build and repair muscle. If protein is eliminated from the diet, then muscle wasting may occur. However, protein can either come from a fatty source or from a lean source. Fatty sources of protein are beef, pork, ham, bacon, high-fat protein shakes, high-fat protein bars, lamb, goat, and so forth. Lean sources of protein include chicken, fish, turkey, egg whites, lentils, beans, and chickpeas. Lentils are rare in the western diet, but are very high in protein and fiber and low in fat, which makes them a lean source of protein. Obviously, eating proteins from lean sources will not increase body fat, but eating proteins from fat sources will increase body fat, so it is better to eat protein from lean sources. Protein is definitely needed in any healthy weight loss. Without it, the body undergoes muscle wasting, which causes the metabolic rate to go down. When the metabolic rate decreases, then fat breakdown is prevented. So increased muscle mass will increase the metabolic rate. The recommended dietary allowance of protein is 0.8g/kg or 56g for a 70kg person. Ten to 25 percent of calories should come from protein. A study done by the University of Illinois McKinley Health Center stated that having 20–25 grams of protein a day is enough to maintain muscle mass.[1] According to a study done by UCLA, about 0.36g of protein per pound of body weight is needed to prevent muscle wasting in the sedentary individual.[2] In athletes, about 0.73–0.82 grams of protein is needed to prevent muscle wasting.

Essential Amino Acids	Nonessential Amino Acids
Histidine	Alanine
Isoleucine	Asparagine
Leucine	Aspartate
Lysine	Aspartic Acid
Methionine	Cysteine
Phenylalanine	Glutamic Acid
Threonine	Glutamine
Tryptophan	Glycine
Valine	Proline
	Tyrosine
	Serine

Table 1. Essential vs. Nonessential Amino Acids

[1] Eitel, J. (2014, January 13) How Much Protein Does a Human Body Need Daily to Maintain Muscle? Accessed from http://www.livestrong.com/article/530295-how-much-protein-intake-to-prevent-muscle-loss/

[2] Eitel, J. (2014, January 13) How Much Protein Does a Human Body Need Daily to Maintain Muscle? Accessed from http://www.livestrong.com/article/530295-how-much-protein-intake-to-prevent-muscle-loss/

The muscle and liver have glycogen stores in them. The muscle uses glucose from these glycogen stores during the initial phases of exercise. After these glycogen stores get depleted, fatty acid from adipose tissue and ketone bodies from the liver are broken down and used as energy by the muscles. This is the concept behind an Atkins diet. It depletes your body of carbohydrates and then it starts to break down the fat. This is a misconception, however. If you are consuming saturated fats and trans fats and breaking down fats, then you are typically breaking even. You are consuming proteins to avoid muscle wasting. However, four to six hours after a meal, your body starts synthesizing its own sugars via gluconeogenesis. Fat cannot be used in gluconeogenesis, because in humans fat cannot be converted into sugars. So gluconeogenesis converts amino acids into sugars. The source of amino acids for breaking down into sugars during gluconeogenesis is muscle protein. Muscle protein is broken down to make this sugar. So, just because you are not consuming sugars does not mean you will start to burn all the fat in your body. The body will break down muscle mass and form sugar, which will be broken down and used as energy. You will have less sugar in your body and you will lose weight, but you are only losing water weight. You are not really losing fat weight, because you are consuming saturated and trans fats in this diet. This is because carbohydrate metabolism and storage requires water. So when you decrease carbohydrate intake, you will lose water. Based on the biochemistry of how the body works, the Atkins Diet does not really decrease body fat. This is definitely not the diet you want to be on if you want to build muscle or lose fat. You cannot stay on this diet and keep your body in a state of ketosis forever. Most people I know have only shown temporary results from this diet; when they get off the diet they typically gain the weight back. This is basically an unhealthy diet to be on. So, in the morning when you wake up after fasting for a while, the body is low on stored carbohydrates. This is why I typically do my aerobic workout in the morning, because after a fasting state the muscle will tap into the adipose tissue for energy a lot faster—so it will burn more fat.

I would say that the South Beach Diet is better than the Atkins Diet. The diet I used has some similarities and also has some differences when compared to the South Beach diet. There is no Oktoberfest exception;

alcohol is to be avoided. The South Beach Diet modifies the types of fats and carbohydrates that are to be consumed, and proteins must come from a lean source. The diet I used also eliminates obesity additives. Another critical factor in my diet is the timing of the meals. Larger meals should be eaten earlier in the day. If you exercise in the morning, meals should be consumed after the exercise. There should be a four-hour fast between your last meal and the time you sleep—no night-time meals. If you exercise at night, then you need to fast four hours before the exercise, and then sleep after the exercise. This is typically the way I would do it. This is because carbohydrate intake before exercise will decrease fat oxidation.[3] This decrease in fat oxidation lasts for four hours; that is why I always fast for four hours before exercise.

Some people feel that fasting is a good way to lose weight. However this is an unhealthy type of weight loss. After 10–18 hours of fasting, liver glycogen stores become depleted. Gluconeogenesis increases as glycogen stores are depleted. As I mentioned before, from 4–6 hours after the last meal—and up to several weeks of fasting—the muscle protein begins to break down and gets turned into glucose via gluconeogenesis by the liver. This process becomes more rapid after the first few days of fasting. The body goes into muscle wasting, which may make a person feel weak. Under no circumstances do you want to undergo muscle wasting during weight loss because this would be an unhealthy weight loss. You will also become vitamin deficient by starving yourself. During a period of starvation, body fat stores are broken down, but muscle mass is also broken down. During a diet, you want to avoid breaking down muscle mass, but you want to deplete the body of its fat stores.

---

[3]   Horowitz, J.F., Mora-Rodriguez, R., Byerley, L.O., Coyle, E.F., (1997, October 1) Lipolytic Suppression Following Carbohydrate Ingestion Limits Fat Oxidation During Exercise. *American Journal of Physiology-Endocrinology and Metabolism*, 273 (4) E768-E775. Accessed from http://ajpendo.physiology.org/content/273/4/E768.full

Chapter 3

How to Determine the Nutritional Value of Food

The most important thing when it comes to weight loss is to know the nutritional value of the food we eat, typically done by reading labels. We have discussed basic nutrition in chapter 2. From this we can conclude that we need to stay away from foods that have a low nutritional value that increase body fat. We need to learn to eat foods with a high nutritional value that decrease body fat. Foods high in monounsaturated fats, polyunsaturated fats, fiber, low glycemic index foods, and protein from lean sources have a high nutritional value and should be consumed. Foods with high amounts of saturated fats, trans fats, simple sugars, starches with low fiber, high glycemic index foods, and fatty sources of protein have a low nutritional value and should be avoided. We must also read the ingredients carefully. If there are less than 0.5 grams of a certain fat, carbohydrate, protein, and so forth in a food, then the FDA does not require the company to put that amount on a food label. One very important thing to note is that fitness experts, health nuts, and bodybuilders all think alike when it comes to food. In order to look like them, you have to think like them—and eat like them.

Here are some examples. We will first compare labels on some protein shakes. Let's look at Protein Shake A, GNC's Amplified Wheybolic Extreme 60 (Table 2) versus Protein Shake B, Muscle Milk (Table 3). We see that Protein Shake A has more protein per serving and Protein Shake

B has more saturated fat, sugar, and fiber per serving. Although Protein Shake B has more fiber, I would still choose Protein Shake A because it has less saturated fat and sugar. There are lots of protein shakes out there that are loaded with fat and sugar. So not all proteins shakes will help you burn fat.

Total Fat	
Saturated Fat	1g
Cholesterol	15mg
Sodium	300mg
Potassium	450mg
Total Carbohydrate	7g
Dietary Fiber	1g
Sugars	2g
Protein	60g

Table 2. Amplified Wheybolic Extreme 60 by GNC, amounts per serving.

Total Fat	12g
Saturated Fat	6g
Trans Fat	0g
Cholesterol	25mg
Sodium	310mg
Potassium	440mg
Total Carbohydrate	18g
Fiber	5g
Sugars	4g
Protein	32g

Table 3. Muscle Milk by Cytosport, amounts per serving.

Let's look at another example. Protein Shake A, Jay Robb Whey Protein (Table 4) versus Protein Shake B, Cytosport's Monster Milk (Table 5). Protein Shake A has less saturated fat and sugar than Protein Shake B. Protein Shake B does have more fiber in it, but I would still go with Protein Shake A. My reason for choosing Protein Shake A is because it has less saturated fat and sugar. You want to avoid saturated fats and sugar if you want to lose weight.

Total Fat	0g
Saturated Fat	0g
Trans Fat	0g
Cholesterol	0mg
Sodium	150mg
Potassium	390mg
Total Carbohydrate	1g
Dietary Fiber	0g
Sugar	0g
Protein	25g

Table 4. Jay Robb Whey Protein, amounts per serving.

Total Fat	9g
Saturated Fat	3g
Cholesterol	15mg
Sodium	430mg
Total Carbohydrate	17g
Dietary Fiber	5g
Sugars	4g
Protein	50g

Table 5. Cytosport's Monster Milk, amounts per serving.

Now, let's compare some protein bars. Many protein bars are disguised as health foods. However, very few protein bars actually have a high nutritional value. Let's examine Protein Bar A, Quest Bars, (Table 6) compared with Protein Bar B, Supreme Protein Bars (Table 7). We see there is a lot more sugar, protein, and saturated fat in Protein Bar B. There is much more fiber and a lot less sugar and saturated fat in Protein Bar A. For this reason, I would choose Protein Bar A.

Total Fat	9g
Saturated	1g
Polyunsaturated	0g
Monounsaturated	0g
Trans	0g
Cholesterol	5mg
Sodium	360mg
Potassium	135mg
Total Carbs	22g
Dietary Fiber	18g
Sugars	1g
Protein	20g

Table 6. Quest Bar, amounts per serving.

Total Fat	8g
Saturated Fat	5g
Trans Fat	0g
Cholesterol	10mg
Sodium	150mg
Potassium	30mg
Total Carbohydrate	18mg
Fibers	1g
Sugars	5g
Other Carbs	14g
Protein	15g

Table 7. Supreme Protein Bar, amounts per serving.

Now let's look at an example where labels can be misleading. The label for Pillsbury Sugar Free Cake Mix says it has no trans fats. In the ingredients list, however, it says there is partially hydrogenated soybean oil, which is a trans fat. This is because there is less than 0.5 grams of trans fats, so the FDA does not require the company to put it on the nutrition label.

What about the nutritional value of foods that do not have labels, such as fruits and vegetables? The nutritional value of fruits and vegetables can be found online. Fruits and vegetables play a major role in any weight loss diet. Most fruits and vegetables are typically high in fiber, which decreases their glycemic index. They are also low in calories. It also takes the body a lot of calories to break down fruits and vegetables. As a matter of fact, some experts even argue that most fruits and vegetables are negative calorie foods. This means that more calories are required to break them down than the number of calories in the food itself. However, this statement is controversial. We do know that most fruits and vegetables are high in water, which has no calories. We also know that many fruits and vegetables are high in fiber, which is not absorbed by the body, and they are low in calories. So based on this, a good weight loss diet would consist of a lot of fruits and vegetables with a low glycemic index.

Another thing worth mentioning here is alcohol. Some people in their twenties and thirties consume large quantities of alcohol. Alcohol is also converted to fat and stored as body fat. Alcohol is also known to increase estrogen levels and decrease testosterone levels. Excess alcohol can even cause male breast growth. When I heard this, I stopped consuming alcohol. No matter what your goal is—whether it is to lose weight or build muscle—alcohol will only hinder in the process.

Now we'll talk about something referred by many as *obesity additives*.[456] Obesity additives are added to foods for various reasons. Obesity additives contribute to obesity and foods with them should be avoided. Some examples of obesity additives are: sodium nitrates, artificial sweeteners, artificial colors, flavoring, monosodium glutamate, caffeine, refined sugar, nitrites, BHA, BHT, partially hydrogenated

---

[4]   Lovitch, M. (2010, September 30) Obesity Additives......the Bane of Processed Foods. Accessed from http://davieshealth.blogspot.com/2010/09/obesity-additives-bane-of-processed.html

[5]   Scotti, K. (2008, January 18) The Dirty Dozen-12 Foods/Food Additives to Avoid and Why? Accessed from http://fooddemocracy.wordpress.com/2008/01/18/the-dirty-dozen-12-foodsfood-additives-to-avoid-and-why/

[6]   Bosch, L.T. (2010, November 23) Top 10 Food Additives to Avoid. Accessed from http://foodmatters.tv/articles-1/top-10-food-additives-to-avoid

vegetable oil, pesticides, brominated vegetable oil, olestra, genetically modified organisms, stripped carbohydrates, excess sodium, high fructose corn syrup, and trans fats. Obesity additives are put into processed foods for various reasons. This is why processed foods should be avoided and natural foods should be consumed. According to a website, obesity additives work in three ways.[7] They interfere with the hormone leptin, which controls satiety, letting us know when we are full. Obesity additives also increase insulin, which causes the body to store the calories we eat as fat. These calories do not get stored as muscle energy, so this increases body fat. Thirdly, obesity additives can be addictive. As the old saying goes, if a caveman did not eat it, then neither should you. You are better off eating foods that you can pick off of a tree rather than eating processed food from a factory.

---

[7]   Lovitch, M. (2010, September 30) Obesity Additives......the Bane of Processed Foods. Accessed from http://davieshealth.blogspot.com/2010/09/obesity-additives-bane-of-processed.html

Chapter 4

Basic Muscle Physiology During Exercise

To build muscle, it's a good idea to know how muscle works. Muscle fibers are made up of small microscopic units called *sarcomeres*. Sarcomeres are made of both thick and thin filaments, and during a muscle contraction, the thin filaments interact with the thick filaments. In order to relax, the muscle needs to have ATP, which is derived from breaking down sugars and fats. In a corpse, the muscle has no ATP, and without ATP the muscle can go into rigor mortis or tetany; the muscle would remain in the contracted condition. ATP is produced by breaking down sugars and fats during both glycolysis and the citric acid cycle. During exercise, adrenaline is released from the adrenal gland and it stimulates the release of glucagon, released from the pancreas. These two hormones increase the amount of glucose and ketone bodies released into the bloodstream. The glucose comes from the liver by breaking down the glycogen stores via glycogenolysis and gluconeogenesis, which we discussed earlier. The muscle proteins are broken down during intense exercise, and the amino acids are used in the process of gluconeogenesis. Ketone bodies form in the liver from breaking down the fat from the adipose tissue, which is made up of adipocytes or fat cells. This lipolytic or fat breakdown process by glucagon is minimal in human beings. Epinephrine and a lack of insulin have more of an effect on fat breakdown in fat cells during exercise. So, during exercise a cascade of events occurs in the liver, causing sugar to increase in the bloodstream. Fatty acid and ketone bodies are also released into the bloodstream. Sugars, ketone bodies, and fatty acids are used as energy during exercise. So, it is good to consume protein after a workout to help build and repair muscle.

Exercise has two effects on muscles.[8] One effect is *hypertrophy*, a process in which the muscle cells enlarge. This happens when someone uses fewer repetitions with high weights. Then there is a process called *hyperplasia*, when muscle cells increase in number. This happens when someone does more repetitions with low weight. This hyperplasia theory is controversial. During the catabolic phase of my transformation, I did more repetitions with low weight. In the anabolic phase of my transformation—in addition to doing more repetitions with low weight—I did fewer repetitions with high weights. So, I was increasing both the size and number of muscle cells.

The two types of muscle contractions that happen in everyday life are *isotonic* and *isometric*. Isotonic contractions involve a change in the muscle length and increase in muscle tension. It is the typical contraction that occurs in the process of lifting a weight. This exercise increases the strength, size, and definition of muscle. There are two categories of isotonic movement: *concentric* and *eccentric*. Concentric movement involves the muscle shortening upon contraction, while eccentric movement involves the muscle lengthening upon contraction. Isometric contractions have a constant tension on the muscle, but the length of the muscle does not change. There is no joint motion in this. This is the contraction that occurs after you lift up a weight and hold it in place. Isometric exercises develop power and sustain strength; they are not good for people with hypertension.

There are two types of muscle fibers: *Type I* and *Type II*. Type I muscle fibers, or red muscle fibers, are the dominant muscle fibers in the body and are also known as *slow muscle fibers*. They contain more mitochondria, which produces more energy for them. They are used for low intensity movements and can be used for a long duration and repetitive movements. They fatigue slowly and are resistant to injury.

---

8   Rockwell, R. (2011, January 2) High Reps and Muscle Hyperplasia. Accessed from http://www.livestrong.com/article/348093-high-reps-muscle-hyperplasia/

These fibers put out very little force. They are the ones targeted with exercise and they grow in response to exercise. They are used when you stand, walk, or lift light objects, and they use fat primarily as a fuel. Type II, or *white muscle fibers*, are fast twitch muscle fibers. They are used during high intensity actions like jumping or lifting heavy weights and they use glucose primarily as a fuel. They are further divided into Type IIA, IIB, and IIC. Type IIA muscle fibers are stronger and larger than Type I muscle fibers; they produce high force for long periods of time. Type IIB are found more in sedentary people. They are very susceptible to injury, and when they are used they are damaged. Typically, when a sedentary person begins an exercise program, these fibers start to go down in number. Type IIC occur when Type IIB combines with satellite cells. These cells are still being studied and not much is known about their properties.

Chapter 5

Body Mass Index, Body Fat Percentage, Lean Body Mass, Ideal Body Weight, Basal Metabolic Rate, and Body Types

W hen dealing with the topic of weight loss and obesity, certain measurements come into play. All of these measurements should be taken in the morning before eating and after having your bowel movement. The first measurement is Body Mass Index or BMI. BMI is measured by taking the weight of the person in kilograms and dividing it by the height in meters Categories of BMI are: underweight, normal, overweight, mildy obese, moderately obese, and severely obese. Bear in mind that BMI does *not* take into consideration how much body fat the person has. A muscular person with a low amount of body fat can be considered obese. (BMI = weight (kg)/(height (m) x height (m).) Below 18.5 is considered underweight. Between 18.5 and 25 is normal weight. Between 25 and 30 is considered overweight. Between 30 and 35 is considered mildly obese. Between 35 and 40 is considered moderately obese. Over 40 is considered severely obese.

The next parameters we will look at are *body fat percentage* and *lean body mass*. Body fat percentage also determines if a person is obese or not (Table 8). However, it does take into consideration the person's body fat. There are many ways to calculate body fat percentage, but the easiest way to do it is by the tape measurement method. For men, the age, height, weight, neck circumference just below the Adam's apple, and the abdominal circumference at the level of the umbilicus are measured and can be plugged in on any website that measures body fat percentage, such as http://www.linear-software.com/online.html. For women, the age, height, weight, neck circumference, abdominal circumference at the narrowest point along the trunk (waist), and the circumference at the widest point (hips) can be plugged into the same website. The lean body mass is basically the weight of the body without the fat. It can easily be calculated by subtracting the body fat percentage from 100, dividing that number by 100, and then multiplying it by the weight. ((100 - BF%)/100) x Body Weight = Lean Body Mass.)

	Women	Men
Essential Fat	10–13%	2–5%
Athletes	14–20%	6–13%
Fitness	21–24%	14–17%
Average	25–31%	18–24%
Obese	32%+	25%+

Table 8. Body Fat Percentage.[9]

Let's talk about Ideal Body Weight. There are many ways to measure the ideal body weight. I typically like the Hamwi method the best.[10] It's another quick parameter to measure whether or not a person is overweight, in addition to BMI. It has stricter criteria. The formulas are summarized below. An estimate of your body frame can be calculated by wrapping your thumb and index finger around your wrist. If they touch then you have a medium frame. If they overlap then you have a small frame. If they do not meet then you have a large frame.

---

[9]   Perry, M. (2010, August 3) Ideal Body Fat Percentage Chart: How Lean Should You Be? http://www.builtlean.com/2010/08/03/ideal-body-fat-percentage-chart/

[10]   Hamwi Formula. Accessed from http://www.nafwa.org/hamwi.php

Hamwi Formula for Men

106 lbs. for 5 feet, and 6 lbs. for every inch more than 5 feet if the person has a medium frame.

If they have a small frame, subtract 10 percent; if they have a large frame, add 10 percent.

Hamwi Formula for Women

100 lbs. for the first 5 feet and 5 lbs. for every inch more than 5 feet (medium frame).

If they have a small frame, subtract 10 percent; if they have a large frame, add 10 percent.

I like to use this method since it's quick and easy. It gives you a good idea of where you stand and how much weight you need to lose to look good. It is stricter than the BMI method. If you are at your ideal body weight, you will be in the normal BMI range. However, you can be within the normal BMI but may not be at your ideal body weight. About two-thirds of the people in America are obese or overweight. Less than one-third would be at their ideal body weight according to this method. Most women would have an attractive figure if they were at their ideal body weight. Models would typically be at least ten pounds below their ideal body weight. It's a little more complicated for men. Most men want to have a muscular figure, and most women want to stay slim. Ideal body weight is a better measurement for women who want to stay slim than it is for men who want to be muscular. Some men could be very muscular and also be significantly above their ideal body weight. This is because muscle is heavier than fat and having more of it can certainly increase your body weight. As a matter of fact, some muscular men could have a BMI that puts them in the obese category, but they would have a low body fat percentage. Some men can have good muscle tone when they are ten pounds above their ideal body weight. So for men who want to be muscular, ideal body weight is not really a good measurement. For the two-thirds of men in America who are overweight or obese with a high body fat percentage, ideal body weight can give them an idea of where they need to be.

Basal Metabolic Rate is the amount of energy the body burns at rest in a day. It can be calculated online on different websites based on height, weight, age, and sex. Here is one website:. http://www.bmi-calculator.net/bmr-calculator/. Below are the formulas used to calculate your basil metabolic rate:[11]

Women: BMR = 655 + (4.35 x weight in pounds) + (4.7 x height in inches) – (4.7 x age in years)

Men: BMR = 65 + (6.23 x weight in pounds) + (12.7 x height in inches) – (6.8 x age in years)

Ellia et al. and Wang etal. described the specific metabolic rates of different organs as follows: "200 for liver, 240 for brain, 440 for heart and kidneys, 13 for skeletal muscle, 4.5 for adipose tissue, and 12 for residual organs and tissues in young and middle aged adults; 194 for liver, 233 for brain, 426 for heart and kidneys, 12.6 for skeletal muscle, 4.4 for adipose tissue, and 11.6 for residual organs and tissues in adults over age 50," in kg/kcal per day.[12] The heart and kidneys have the highest metabolic rates. It would make sense that if you raised your heart rate by doing aerobic exercises, then you would raise the metabolic rate of your heart. In addition, aerobic exercise would raise the metabolic rate of your skeletal muscles. Both of these factors combined will cause you to burn more fat. This is why aerobic exercise can enhance your weight loss when used in combination with diet.

When looking at global obesity rates, we see that the obesity and overweight rates in America are much higher than in other countries when using the BMI method.[13] In 2010, about one-third of the population

---

[11] BMR Calculation. Accessed from http://www.bmi-calculator.net/bmr-calculator/bmr-formula.php

[12] Wang, Z., Ying, z., Bosy-Westphal, A., Zhang, J., Schautz, B., Later, W., HeymsField, S.B., and Muller, M.J. (2010 December) Specific Metabolic Rates of Major Organs and Tissues Across Adulthood: Evaluation by Mechanistic Model of Resting Energy Expenditure. *American Journal of Clinical Nutrition.* 92(6): 1369-77. Accessed from http://www.ncbi.nlm.nih.gov/pubmed/20962155

[13] OECD Health Data 2005. Accessed from http://www.oecd.org/newsroom/35625122.pdf

was obese and two-thirds were overweight. The numbers were about the same in 2000. In 1990, the rates of obesity and overweight people were one-fourth and one-half, respectively. This epidemic started around 1987. This has a lot to do with the low nutritional value of the food being eaten here. Many experts will say that American food contains lots of obesity additives, which we talked about before. Excessive drinking and a sedentary lifestyle contribute to obesity, but most of it has to do with the food we eat. I travel to Europe a lot and when I do I typically lose about five pounds, which I gain back when I come home. I typically do not exercise when I am in Europe. So, based on my own personal experience, how we eat has a lot to do with our obesity rates.

Let's talk about the three basic body types as described by scientist W.H. Sheldon. They are called *mesomorph*, *ectomorph*, and *endomorph*. A mesomorph is a person with a muscular type of body, an ectomorph has a slim or linear body, and an endomorph has a fat or round body. People tend to display characteristics of all three body types, but they are typically dominant in one. This based on a three-digit rating system. Each digit is given a seven-point rating. The first digit signifies endomorphy, the second one mesomorphy, and the third one ectomorphy. The person who is 171 is an extreme mesomorph. Someone rated 711 is an extreme endomorph, someone rated 117 is an extreme ectomorph. Most people would fall between 333 and 444.

Chapter 6

How to Get Rid of the Pseudogynecomastia, Love Handles, and the Abdominal Fat and Replace Them With a Toned chest and Six-pack Abs

Men often have the appearance of breasts. The medical terminology for this is *gynecomastia*, male breasts with actual breast tissue. This is very rare and is found in certain disorders. More commonly, men have *pseudogynecomastia* or "false breasts." This is the appearance of breasts but without real breast tissue. They are composed of fat.

Pseudogynecomastia typically develops after puberty, primarily because that is when the testes release testosterone. Around 40 to 60 percent of the male population has pseudogynecomastia. Some of this testosterone gets converted into estrogen by an enzyme called *aromatase*. Aromatase is an enzyme found in the fat cells of the body. Typically, when there are more fat cells in the body, more estrogen is produced. This leads to pseudogynecomastia. So, the first thing to do to get rid of the fat cells would be to lose weight.

Certain things in the diet can cause pseudogynecomastia. If someone consumes too much alcohol, this may contribute to it. Alcohol is known to increase estrogen levels and decrease testosterone levels. So it would be a good idea to stop alcohol consumption. Believe it or not, milk contains a lot of estrogen. So it might be a good idea not to drink too much. Whole milk contains a lot of saturated fat as well. In my diet I did not drink anything other than water, decaffeinated green tea, and lean protein drinks. Marijuana is also known to decrease testosterone and increase estrogen. Some argue that soy protein may also increase estrogen levels, but this is controversial. Logically, any food that increases body fat would contribute to pseudogynecomastia.

Eating certain foods helps get rid of pseudogynecomastia. Cruciferous vegetables contain *indole 3 carbinole*, which is known to decrease estrogen levels and increase testosterone levels. I recommend eating cruciferous vegetables; it is best to steam them or eat them raw. Some vegetables that contain indole 3 carbinole are: broccoli, cauliflower, radish, watercress, Brussels sprouts, cabbage, kale, and collard greens. Zinc also helps get rid of male breasts because it increases testosterone levels. Foods with zinc include lamb, pork, beef, dark chicken, peanuts, oysters. At the same time, it is not a good idea to eat lamb, pork, or beef, because they are also known to be fatty meats. Fatty diets must be avoided in order to suppress pseudogynecomastia. Calcium is known to regulate estrogen uptake and can also be used against pseudogynecomastia. Avocados also help get rid of pseudogynecomastia. They are rich in monousaturated fat, which helps in the production of testosterone. Vitamin B is also helpful. Again, the most important thing here is to decrease the fat cells by losing weight. In severe cases, or when all else fails, surgery may be needed.

It is everyone's dream to get rid of abdominal fat and get the six-pack abs, so they can look like someone on a magazine cover. Typically, people with six-pack abs have more muscle tone. In order to have more muscle tone, you would have to bring down your body fat percentage. How can anyone see muscle tone if all of the fat surrounds the muscle? Typically, you would want a body fat percentage of less than 8–12% in order to see six-pack abs, if you are a man. My abs first became visible at a body fat measure of 11 percent and 140 lbs. For women, abs start to be visible sooner, at less than 15–17% body fat. This is because women have a tendency to store fat around the buttocks and thighs. The abdomen is the hardest part of the body to tone up for men. This is because men have tendency to store fat around the abdomen. When the body starts to accumulate fat, it typically starts to accumulate more around the abdomen than the rest of the body. When the fat starts to come off, it typically starts to come off the body overall, and the last place left with fat is the abdomen. Further dieting can get rid of this residual fat. Those with six-pack abs generally have good muscle tone. I have seen people who have muscle tone all over the body, but not in the abdomen. They would have to diet further to get muscle tone in the abdomen. I have never seen anyone with muscle tone in the abdomen who didn't also have it all over the body. Once you have reached the stage where you have good muscle tone in the abdomen, you will have good muscle tone everywhere. The lower abdomen is the last place where the muscle tone develops in men. The area around and below the umbilicus is where the fat will come off last in men.

A common misconception that most people have is the possibility of spot reduction. Many people believe that by doing sit-ups or other ab exercises, their abs will start coming out. This may strengthen the abs, but it will not get rid of the abdominal fat. I have done enough abdominal surgeries to know that underneath the skin there is a layer of fat. Underneath the fat is the *fascia*, which encases the ab muscles. In order to make the ab muscles visible, you would have to decrease or get rid of the fat layer. Love handles are nothing more than fat stores along the side of the body. When the body does not have any more space to store the fat, it starts to store it in the love handles and also in the chest. This also makes pseudogynecomastia more prominent. During weight loss, the love handles disappear before the belly does. There is not an exercise in the world that will make love handles disappear. Only dieting will make love handles go away. In some skinny people who have little body fat, the abs may not be evident. This is because the person may not exercise the abs, so they are not visible. So to get the six-pack abs, you need a combination of diet and exercise: 95 percent diet and 5 percent exercise. Unless you are a very thin person with a very low body fat percentage, you cannot get visible abs by doing ab exercises. If you are like most people and have some belly fat, then you would have to get rid of the belly fat first before you can even see the abs. If, after getting rid of the belly fat, the abs are still not visible, then ab exercises are needed to make them visible. You will find many ads on the Internet that claim you can get six-pack abs in a very short period of time just by doing a certain exercise or following some diet. Of course, when many people try to do this, they fail. The point is that if you are a really obese person, it is physiologically impossible for you to get six-pack abs quickly. People who have little abdominal fat can get six-pack abs in a short period of time, but it's highly unlikely if you have a lot of abdominal fat. A person with more abdominal fat would need to diet longer to get them. As a previously obese person, it took me five months of dieting and some aerobic exercise before I even began to see my abs at a weight of 140 lbs. and body fat percentage of 11 percent. I used to do a lot of ab exercises before then. My abs were already built up, but I could not see them with all of the fat there. So, if you are an obese person, you can still get six-pack abs, but it will take you longer.

So, to summarize, the key to a toned chest and six-pack abs is weight loss. The key to weight loss is a good diet. Exercise can enhance a good diet, but cannot be a substitute for it. There is no substitute for a good diet.

Chapter 7

Total Body Transformation

Total body transformation is the process in which you change your unhealthy, overweight, obese body into a healthy, muscular, lean, slender toned body. Total body transformation has two phases. The first phase is the *catabolic phase*, when you break down all of the body fat. The second phase is the *anabolic phase*, when you build up muscle mass. Not everyone undergoes goes both phases of the transformation; it all depends on your goals. Since men typically want to burn fat and build muscle, they would undergo both phases. Women typically want to maintain a slim figure and burn fat, so they would only need the catabolic phase. However, some men are only interested in burning fat and therefore only need the catabolic phase. If a man is very thin and has a low body fat percentage, then he would not need the catabolic phase. For the average man, it wouldn't make sense to do only the anabolic phase. He would retain body fat and build up the muscle underneath. It would not make him look more attractive or make him healthier. Women who want to burn fat and build muscle would need both phases. Then, of course, there are thin men who want to gain muscle mass. They would need to just undergo the anabolic phase. Once again, before starting a total body transformation, always check with your doctor.

The catabolic phase comes first and consists primarily of dieting, the fastest way to take off the body fat. Aerobic exercise can enhance the catabolic phase. This phase involves a low-calorie diet with fruits and vegetables that are high in fiber and a small portion of protein with each meal to avoid muscle wasting. The workout for the catabolic phase would be lower tension or weights and more repetitions. It would be a continuous workout in which one muscle group is worked out right after another without stopping; this would make it most effective. I like to do my aerobic exercise in the morning before eating. This forces the body to tap into the fat that you are trying to get rid of and use it as energy. The body starts to lose fat all over and not just in one area. Fat loss occurs from the periphery to the core. This means that the arms, legs, love handles, and so forth will lose fat first, and then fat loss will occur in the core of the body.

A common mistake people make when it comes to weight loss is that they start an exercise program, but they continue to eat foods that increase body fat. After a few weeks of no results, they lose hope and stop exercising. They then resume their old lifestyle. Let's look at this in a rational manner. If you are exercising, then you are burning fat. If you are continuing to eat foods that increase body fat with that exercise, then you are more or less breaking even. You are pretty much just cycling the fat. You can change your diet and eat only foods that will *not* increase your body fat, and then you will begin to lose the weight. If you enhance that diet with aerobic exercise, then you will lose weight at a faster pace. The question comes up, *How much weight do I need to lose in my catabolic phase?* The answer depends on what your goal is; it depends on what makes you happy when you look in the mirror. You can lose as much weight as you want as long as you maintain a BMI and body fat percentage that is within a normal healthy range. You should never be at a BMI or body fat percentage that is unhealthy. This can lead to other problems.

The body does use fat for many purposes. The body uses fat for physical protection. It pads bony surfaces and uses it as a source of energy. The brain and nervous system need fat for normal functioning. Fats are also a good source of fat-soluble vitamins. They also line cell membranes, protect against toxins, and provide strength to the bones. Fats provide the body with immunity and also assist in the lungs with respiration. They also make skin more resilient. So the body does need fat, and it is not a good idea to go down too low on body fat.

At the end of my catabolic phase, I reached a BMI of 18.4, which is just below the cutoff of 18.5 for what is healthy. I would not recommend that anyone else do this. I put that weight back on during my anabolic phase in the form of muscle weight. If at any point during your catabolic phase you feel weak, dizzy, or feel like you might faint, then it is not a good idea to lose any more weight. At this point the catabolic phase needs to stop. One thing to note is that if you are a hypertensive on medication, you may feel dizzy at some point during your catabolic phase. This is because weight loss brings down blood pressure. This is what happened to me. This is one reason it is very important to discuss your weight loss plan with your doctor. If your blood pressure goes down during the catabolic phase of your transformation, your doctor may need to discontinue or lower the dosage of your blood pressure medication. You should never discontinue or change the dose of your hypertensive medication without consulting with your physician. At the end of my catabolic phase, I could see vascularity in places I had never seen it before. I had vascularity in the lower and upper abdominal muscles as well as on the oblique muscles, which go all the way up to the axilla on both sides. According to the website thisiswhyyourjacked.com, I would be under the category called *shredded*, and my body fat measurement would be between 4–7 percent. This is consistent with my measured body fat percentage of 4.4 percent at the time. The only category lower than shredded is *sliced*. The body fat measurement here is 3 percent. This is not a healthy state and can only be sustained for a few hours.

As time goes by, weight becomes harder to take off. If your plan is to take off a lot of weight, then you may encounter weight loss plateaus in the catabolic phase. I had two weight loss plateaus during the course my weight loss. The first one was at 150 pounds. I overcame this by adding a three-mile run to my routine. Then at 130 pounds I encountered another weight loss plateau, even with the three-mile run. Eating less food at this point did not help me lose weight, so I decided to eat more fruits and vegetables. I then got through this plateau.

One big problem you may run into after weight loss is loose skin. This depends on your age, the amount of weight loss, and how fast the weight was lost. At times, the skin can just readjust to your new body, but it may take some time. Some loose skin may remain after the catabolic phase, but when you build up muscle in the anabolic phase it might go away. I had some loose skin in the abdominal area. I used lotions with collagen and elastin to overcome this problem. Eventually it started to look better, although not perfect. In severe cases, surgery may be required.

The anabolic phase consists of strength training by weights or tension. This phase follows the catabolic phase. In this diet, more protein is consumed. The protein should come from lean sources, not from fatty sources. The workout for this phase would be less repetition and more tension or weights. You should always start out with less tension and weight and gradually work your way up; otherwise you put yourself at risk for injury. Typically, I start at a tension on the Bowflex where I can do ten repetitions. If I can do more than ten repetitions, then I need to increase the tension. If I can't do more than ten, then I need to decrease the tension. You do run the risk of injury when doing any type of workout, whether it is aerobic or strength training. If an injury occurs, then it is best to stop working out and let the injury heal. Once the injury is healed, then you can start working out again.

The physiology behind muscle building is straightforward and simple. When a muscle group works against higher weights or tension, the cells begins to undergo *hypertrophy*; in laymen's terms, they get larger. So, working out a certain area of the body will not cause that area to lose fat, but working out a certain muscle group against high tension or high weight will cause it to get larger. So from this we can conclude that fat reduction does not work the same as muscle gain. The question comes up, *How much muscle do I need to gain?* Again, it depends on what makes you happy. Your genetics is only going to allow you to gain a certain amount of muscle. For some people that will be a little. However, this does not mean that you cannot look good. It is not really the muscle mass that makes people look good; it is good muscle tone. When you burn off enough fat, then good muscle tone will make the muscles more pronounced. You will look more muscular with good muscle tone. You should never try to go above your body's natural ability to gain muscles by taking steroids. This may cause you to gain some muscle mass, but it will not make you to look much better than you did if you just had good muscle tone. Also, the use of steroids has been associated with gynecomastia. This is because some the external testosterone that is injected into the body will be converted into estrogen by the *aromatase* enzyme. With all of the unhealthy effects of steroids, it is just not worth it.

It is ideal to begin strength training in the anabolic phase after your catabolic phase. Testosterone levels in men increase as the fat is burned off, because there is less *aromatase*. Aromatase is the enzyme found in fat cells that converts testosterone into estrogen. As you burn off the fat there is less aromatase. Aromatase is also found in the granulosa cells of the ovaries, and it is what produces most of the estrogen in females. This testosterone effect is significant if a lot of fat is burned off. It is like being on a continuous mini-dose of steroids. However, instead of using an exogenous source of testosterone, the body utilizes its own endogenous source of testosterone. Theoretically, you should be able to lift more, do more repetitions, run faster, and so forth. In my case, I lost 81 pounds during my catabolic phase, a significant amount of fat loss. I built up 13 pounds of muscle in just two weeks; the two reasons for this are the testosterone effect and the fact that I had worked with high resistances before. So, if this testosterone effect is combined with a high protein diet and also strength training, then muscle can be built up fast.

One negative side effect of testosterone is that it has an unfavorable effect on the lipid profile. It increases LDL and decreases HDL. This can increase the risk of cardiovascular disease. However, that didn't happen in my case because my HDL went up and my LDL went down. The total body transformation works around this negative side effect. I raised my testosterone levels and also brought my body fat percentage down. Bringing my body fat percentage down helped get me the favorable lipid profile, and it superseded the negative effects of the testosterone on my lipid profile. In short, I managed to get the benefits of the excess testosterone, but I managed to blunt the negative effects of testosterone on the lipid profile.

The anabolic phase is not for everyone. If you are weak, in bad health, or have any injuries, then the anabolic phase is not recommended for you. Always consult with your physician before beginning any diet and exercise plan. I mentioned that diet is the most critical thing in the catabolic phase. In the anabolic phase, consuming a high protein diet is critical. However, it is also critical to strength train in the anabolic phase. Working muscle against high tension or weight is how to make them grow. Not everyone is in the physical shape to do so. Just because you are not in the physical shape to do the anabolic phase does not mean you cannot look good. Remember, it is muscle tone that makes people look better, not the muscle mass. Muscle tone is primarily obtained in the catabolic phase, so that phase is more critical here. Most of the work done to achieve six-pack abs occurs during the catabolic phase, and most of the health benefits from the total body transformation are from this phase. You get more health benefits from burning off fat than from building up muscle. Losing fat will decrease your chances of heart disease, hypertension, stroke, diabetes, backache, snoring, and so forth. Building up muscle will make you stronger, help prevent you from gaining fat, and increase your metabolism. This result is negligible, as we will soon discuss. Since sugar is stored in muscle, you will have more room to store excessive sugar. Less of the sugar you consume will be turned into fat. However, from a medical standpoint, it is more important to lose fat than to build muscle, because it prevents many chronic illnesses that come with age.

So, basically, the catabolic phase is when you break down the fat and the anabolic phase is when you build up the muscle. Now let's apply this to a real life scenario. Let's suppose you are looking at someone on a magazine cover and you want to look like that person. You would have to match this person's body fat percentage and BMI. You would use the catabolic phase to bring your body fat percentage down to that person's body fat level. Then you would use the anabolic phase to bring your BMI back up to that person's BMI. Understand that this is not always going to be realistic, because many people on magazine covers have a body fat level of less than 4 percent, and this is not a sustainable condition. You can bring in down to 4 percent so that it is sustainable, but not lower than that. Also understand that you may not be able to bring your BMI up to that person's BMI, because your genetics may not allow you to gain enough muscle for that. Again, many people on magazine covers take steroids, which is something that you should not do to bring your BMI up. Also understand that a lot of pictures on magazine covers are airbrushed to perfection. When I did my total body transformation, it was not my intention to look like any particular person. I just wanted to be at the lowest body fat percentage I could be at—around 4 percent—and gain as much muscle as my genetics would allow me to.

An article written by Casazza et al. points out some myths about obesity.[14] The first myth they talk about is that small changes in energy intake and expenditure will produce long-term weight changes. I would definitely agree that this is a myth. I would say from experience that the biggest factors that will influence your weight are what you eat and when you eat it. Energy expenditure and exercise have very little to do with weight loss, though they can speed up your weight loss. Many people who exercise and don't eat a proper diet will see minimal results with their exercise, and then they will stop exercising. The second myth they talk about is that you must set realistic goals for weight loss. I would agree that is a myth. A motivated and disciplined person can set even unrealistic goals and still meet them. I was in the obese BMI category and had a body fat percentage of 27, and I went down to a normal BMI with a body fat percentage of four. This was an unrealistic goal, but I still achieved it. The third myth is that slower weight loss has a better outcome than rapid weight loss. I would say this is also a myth in terms of the weight loss aspect. I lost weight rapidly and managed to keep it off. The big problem you can run into with rapid weight loss is loose skin. When you lose weight fast you are not giving the skin time to catch up with the weight loss, but when you lose weight slowly you give the skin more time to catch up with the weight loss. Eventually the skin usually catches up with the shape of the body after weight loss, but it may take years. Sometimes surgery may be necessary. The fourth myth they talk about is that readiness is not important when it comes to weight loss, but I don't think this is a myth at all. The important factors determining success in weight reduction are the person's readiness, discipline, and motivation to lose weight. The fifth myth is that physical education reduces obesity. I would say that this is definitely a myth. As I have

---

[14]   Casazza, K., Fontaine, K.R., Astrup, A., et al. (2013, January 31) Myths, Presumptions, and Facts About Obesity. *The New England Journal of Medicine.* 368:446-454. Accessed from http://www.nejm.org/doi/full/10.1056/NEJMsa1208051

mentioned before, physical education and physical activity have little influence on obesity; diet matters more.

Then Casazza et al. go into some presumptions. The first presumption is that breakfast prevents people from getting overweight and obese, but I disagree with this statement. It is true the timing of your meals is important, and it is important to eat breakfast. However, it also depends on *what* you are eating. It is better to eat earlier in the day and then leave a four-hour gap between your last meal and when you go to bed. Another presumption is that early childhood habits will influence your weight. I do not agree with this either. Weight is more influenced by the diet you consume as an adult.

Another presumption is that eating fruits and vegetables will result in weight loss and less weight gain. I disagree with that, because it also depends on what else you are eating and when you are eating. If you eat fruits and vegetables and then eat a lot of unhealthy foods that will cause you to increase your body fat, then the fruits and vegetables will not help you. If you eat fruits and vegetables and also eat nutritious food with the right eating pattern, then you can lose weight.

Another presumption is that snacking contributes to weight gain and obesity. Again, it depends on what you snack on. Is your snack nutritious? Nutritious snacks will not cause weight gain, whereas non-nutritious snacks can cause weight gain. The timing of the snacks also makes a difference; midnight snacks and snacks less than four hours before bedtime can cause weight gain. Another presumption that isn't true is that the built-in environment—such as having a sidewalk and a park available—will influence obesity.

Then the article goes over some facts about obesity. The first fact is that obesity is genetically influenced, but environmental changes can promote weight loss. It also says that if we identify the environmental factors, we can influence obesity. I have already identified the environmental factor and it is the American diet. The food has a lot of obesity additives. People have a tendency to eat a lot of non-nutritious food, and many people in their twenties drink a lot of alcohol. The timing of the meals is not right either. Many people skip breakfast, have a big dinner, and eat before going to sleep. These are the environmental factors that cause obesity. The second fact is that diet (meaning less food intake) is effective in reducing obesity, but trying to get people on diets does not generally work well. People cannot live off diets on which they are starving, so these diets are sure to fail. For this reason, it is better to be on diets with low calorie foods (or negative calorie foods) and lean protein sources. This way dieters can continue to eat, but they are eating foods that will cause weight loss and still maintain muscle mass. Whether or not one can stay on this diet will depend on the motivation, readiness, and discipline of the individual. Another statement they make is that exercise can improve one's health, even if there is no weight loss. I can imagine that a person who does exercise is better off than the person who doesn't. However a person who exercises and loses weight is far better off than someone who exercises and has no weight loss.

Another statement is that high levels of physical activity help with weight management. This is true, but most people do not have the time to perform these high levels of physical activity. However, diet is still more important. Let's suppose that someone participates in high levels of physical activity, but continues to eat foods that will increase body fat. They will have trouble losing the fat. Many obese people are not able to perform high levels of physical activity. So what do you do for these people? They would have to alter their diet to achieve weight loss. Dieting is a much faster and more efficient way to lose weight than exercise.

Another point in the article is that continuation of conditions that promote weight loss will promote weight loss. This is absolutely true. In order to lose weight, you have to alter your diet and you have to *maintain* that diet in order to keep the weight off. The reason people start to gain weight back after a diet is that they tend to go back on their unhealthy diet. Losing weight and keeping it off is a lifetime commitment. They make another statement in this article. For children, programs that involve the parents and the home setting will promote weight loss. This is true because the parents decide what food the child will be eating. If the parents provide the child with nutritious food, then obviously the child will lose weight. They also say that provision of meals and meal replacement products help with weight loss. This is true if you are eating the right meals and the right meal replacement products. It will have more effect than any holistic approach. They also state that some drugs can help a patient achieve and maintain weight loss. If the person is on a bad diet, then these drugs may not work. If a person is on a medication that burns body fat but they are eating foods that will increase body fat, then they will not lose weight. However, if that person is put on the proper diet, then he or she will not need these drugs and supplements to lose weight. If you are eating a diet that will not increase your body fat, then your basal metabolic rate will burn off the fat. This will cause a net fat loss and you will not need the drug or supplement. The article states that for severely obese patients, bariatric surgery will maintain long-term weight loss and decrease morbidity. This is true, because many severely obese patients will fail on diets. They may not have the discipline, motivation, or readiness.

There are many personal trainers and fitness experts who feel that burning fat and building muscle can be done at the same time. Based on my own experience—and the opinions of many fitness trainers—I strongly disagree with that. Many people argue that by building up muscle, your metabolism goes up and you burn the fat away. When I first brought my Bowflex, I decided to strength train. I built up muscle and went from bench pressing, chest flying, and shoulder pressing 80 to 410 pounds, but I could not lose any fat. The truth of the matter is that very little fat is burned by muscle gain.

Ron Brown, a fitness trainer and author of *The Body Fat Guide*, explains this very well on his website.[15] One pounds of lean muscle will burn 13 calories a day. If someone were to build up one pound of muscle per week, then after ten weeks he or she would have ten pounds of muscle. After ten weeks, the person would burn 130 calories a day, assuming the caloric intake wasn't increased by 130 calories a day or more. So, in 27 days the person would be able to lose one pound of body fat. Ten weeks plus 27 days is about 14 weeks. So, in 14 weeks this person would lose one pound of fat, assuming he or she did not gain any fat while building muscle. During the catabolic phase of my transformation I was losing about ten pounds of fat in one month.

In an article called "Reducing Body Fat Percentage By Gaining Muscle" on the Internet, they compare the results of muscle gain versus

---

[15]   Brown, R. (2002, June 14) Does Building Muscle Burn Away Fat? Hardly! *The Body of Fat Review.* Accessed from http://www.bodyfatguide.com/BuildMuscleBurnFat.htm

fat loss on body fat percentage.[16] Let's assume a man is 170 pounds with 15 percent body fat. His lean body mass is 145 and he has 25 pounds of fat. Table 9 illustrates that after gaining 40 pounds of muscle, his body fat level will go from 15 percent to 11.9 percent. This is a difference of only 3.1 percent. This is also assuming that he does not gain any fat while building up muscle. We are also not taking into consideration the amount of time it will take to add 40 pounds of muscle. For every five pounds of muscle gained, the body fat measurement decreases by 0.4 percent. Again this is a negligible number.

Muscle Gain	Fat Mass	Total Weight	Body Fat Percentage
5 lbs.	25 lbs.	175 lbs.	14.2%
10 lbs.	25 lbs.	180 lbs.	13.8%
15 lbs.	25 lbs.	185 lbs.	13.5%
20 lbs.	25 lbs.	190 lbs.	13.1%
40 lbs.	25 lbs.	210 lbs.	11.9%

Table 9. Impact of Muscle Gain on Body Fat Percentage.[17]

The same article looks at the impact of fat loss on body fat percentage. Table 10 demonstrates that the same 170-pound man went from a body fat measurement of 15 percent to 3.3 percent by losing 20 pounds of fat.

Fat Loss	Fat Mass	Total Weight	Body Fat Percentage
5 lbs.	20 lbs.	165 lbs.	12%
10 lbs.	15 lbs.	160 lbs.	9.3%
15 lbs.	10 lbs.	155 lbs.	6.4%
20 lbs.	5 lbs.	150 lbs.	3.3%

Table 10. Impact of Fat Loss on Body Fat Percentage.[18]

---

[16] Body Recomposition (2008) Reducing Body Fat Percentage by Gaining Muscle-Q and A. Accessed from http://www.bodyrecomposition.com/fat-loss/reducing-body-fat-percentage-by-gaining-muscle-qa.html

[17] Body Recomposition (2008) Reducing Body Fat Percentage by Gaining Muscle-Q and A. Accessed from http://www.bodyrecomposition.com/fat-loss/reducing-body-fat-percentage-by-gaining-muscle-qa.html

[18] Body Recomposition (2008) Reducing Body Fat Percentage by Gaining

The bottom line is that muscle gain is associated with a minimal body fat loss. This is based on the above information and also on my own personal experience. The fastest way to take off fat in large amounts is by diet. This can be enhanced by aerobic exercise. While many experts feel that you can burn fat and gain muscle at the same time, I strongly disagree with this. Burning fat and building muscles involve two different diets and two different types of workout as mentioned above. So, how can it be done at the same time? The more efficient and effective way to burn fat and build muscle would be to do them separately. This is why I feel that a total body transformation is best done with a catabolic phase followed by an anabolic phase. Many body builders have the misconception that before building up muscle mass one must gain weight. This is also wrong. In other words, they feel it is best to gain fat and then build up muscle. Why would you want to add more fat to your body before you build up muscle? This would increase your body fat percentage and make you look less toned and less attractive.

Muscle-Q and A. Accessed from http://www.bodyrecomposition.com/fat-loss/reducing-body-fat-percentage-by-gaining-muscle-qa.html

Chapter 8

Diets For Total Body Transformation

will now discuss my diet for the catabolic phase of my transformation. Once again, before you begin any diet please consult with your physician. My eating and drinking pattern is described below. This is all based on my personal experience, but it did produce results for me. I never really drank anything other than water, decaffeinated green tea, and lean protein drinks. Many drinks are loaded with sugar. Even a glass of orange juice has a lot of sugar in it. I had four cups of green tea a day, two with each meal. The same diet should not to be used every day, because after two to three days you will no longer lose weight. I altered my diet frequently. I had dinner between 6 and 7 p.m. and never ate anything after then. After my transformation, I learned that having my last meal at 5 p.m. did more for me in terms of maintaining my weight. So, I typically did a four-hour fast before sleeping or nighttime exercise. I usually had Amplified Wheybolic Extreme 60 in the morning after working out, a Quest Protein Bar two hours after lunch, and a MHP Power Pak Protein Pudding two hours after that. Sometimes I would sprinkle a little cinnamon in the pudding. This would give me my protein intake each day so as to avoid muscle wasting. My lunch and dinner would change each day.

Here are some examples of my lunch and dinner. For lunch I would have a stalk of celery, one cucumber with the skin peeled with black pepper, half a pack of strawberries, half a pack of raspberries, half a pack of blackberries, and half a pack of blueberries. I would eat Walden

Farm's calorie-free blue cheese dressing with the celery. For dinner I would have the same thing as lunch. Sometimes I would substitute the cucumber and celery with two apples. I would put Walden Farm's calorie-free caramel dip with cinnamon on the apples. I would also use Walden Farm's calorie-free chocolate dip with cinnamon on the strawberries. Another menu I followed was: two crowns of steamed broccoli with either a half a pound of spicy chicken kabobs, salmon, codfish, or spicy chicken burgers. This would be lunch and dinner. One important thing to note is that I always either grilled the meat or baked it. I never fried any of my dishes. It is not good to eat fried food. Some days I would have a serving of lentils, chickpeas, kidney beans, or spinach and mushrooms for both lunch and dinner. Other days I would have a pack of strawberries, blueberries, raspberries, and blackberries for lunch. Then for dinner I would have spinach, kidney beans, chickpeas, or cauliflower. After my total body transformation, I started to add in medium glycemic index foods. Some days I would have three fruit containers that included mango, pineapple, strawberries, blueberries, kiwi, cantaloupe, and a lean protein shake. I would eat this three times a day. Sometimes I would eat this twice a day and eat a grilled piece of salmon. This diet consisted of a lot of low-calorie carbs. I also ate peaches, nectarines, plums, and occasionally apricots. I would also eat nuts once in a while, but not too much. I ate peanuts, cashews, and walnuts. This was the diet that was right for me. However, it may not work for everyone. Choosing the right diet for you takes trial and error. You have to see what you lose weight with. The most efficient way to choose a diet for someone is to have them write down the typical things they eat. Then take out the things that are preventing them from losing the weight and add in healthy alternatives. The diet should not disrupt the person's normal routine too much. If it does, then the diet might not work. The bottom line is that the diet has to be tailored to the individual and not the other way around.

The Atkins diet says we should not eat carbs. As you can see, the above weight loss diet consists of a lot of carbs. As a matter of fact, most of the foods in these diets are carbs, but they are complex low-calorie carbs with lots of fiber. The Atkins diet says carbs need to be completely avoided. However, saturated fats and trans fats are allowed.

This diminishes the nutritional value of the food you eat. No one can live off the Atkins diet for the rest of his or her life. When you come off the Atkins diet, you begin to gain the weight back. It is a short-lived diet. However, the diet I followed is one that people can be on for the rest of their lives. Don't forget that losing weight and keeping it off is a lifetime commitment. It is not a commitment for only a few months. If someone goes on any type of diet and then resumes eating the way they did before the diet, he or she will soon gain back all of the weight. This is very common. The diet I used can be used as a more permanent diet and can get long-term results, which is what people want.

My diet changes in the anabolic phase of the transformation. During my anabolic phase, I was eating more. I mostly ate foods that were high in protein and fiber. I was still avoiding sugars, saturated fats, trans fats, and alcohols. The parameter I kept a close eye on was my abdominal circumference at the level of the umbilicus. I wanted to keep it at 27 inches. As long as it was at 27, my body fat percentage would remain the same. If my neck circumference increased, then my body fat percentage would decrease. I was more interested in building up muscle in this phase, so my weight was less significant to me. As long as my abdominal circumference remained the same, any weight I gained would not be due to fat gain. I consumed more protein during this phase, and it always came from a lean source. I ate chicken, turkey, fish, lentils, chickpeas, kidney beans, protein shakes, protein bars, protein pudding, and so on. Lentils have a very high protein-to-fat ratio. I still ate some carbs rich in fiber with the protein, because the fiber would decrease the absorption of fat. Different types of berries, salads, broccoli, and so forth can be consumed with these proteins. The idea is to maintain a low body fat percentage, thus maintaining a lean, sculpted figure and also increasing muscle mass. This is why I would not recommend eating too much meat from fatty sources, such as pork, beef, lamb, or goat. By putting on muscle you can also put on more fat. So, to minimize the fat that you put on, you would have to consume less saturated fats, trans fats, high glycemic sugars, and alcohol.

Many weight loss experts say that eating small frequent meals is what is needed to burn fat. However, this is very controversial. Megan

McRory's article in the *Journal of Nutrition* contradicts this.[19] The University of Ottowa did a study where they divided people into two groups.[20] Both were given the same caloric intake. One group ate six small meals a day, and the other ate three large meals a day. They both lost about 5 percent of their body weight, so there is research to negate that statement.

19   Reinagel, M., (2011, May 24) A Chink in the Small Frequent Meal Theory. Accessed from http://nutritionovereasy.com/2011/05/a-chink-in-the-small-frequent-meal-theory/

20   Leong, K. (2010, March 31) Can Eating Small Frequent Meals Help You Lose Weight? Accessed from http://healthmad.com/weight-loss/can-eating-small-frequent-meals-help-you-lose-weight/

Chapter 9

Workouts for Total Body Transformation

I had a certain set of exercises that I did in my catabolic phase and then in my anabolic phase. Exercise should always be adjusted to the tolerance level of the individual. While some people may be able to tolerate my level of exercise, others may only be able to tolerate a one-mile walk or a bike ride. Some people may not be able to tolerate exercise at all. However, these people can still undergo the catabolic phase of the total body transformation, working with diet only. The process will be slower. You should always check with your doctor before beginning any exercise program. The same exercises may not always work for different individuals. The exercises described below worked for me, but they may not work for everyone. To find out which exercise is best for you is pretty much trial and error. During the transformation, I would also recommend you monitor your weight, body fat percentage, and blood pressure at least daily.

During the catabolic phase I did mostly aerobic exercises; I did not do any strength training in this phase. I ran for three miles and did 60 push-ups and 76 sit-ups. Most of my exercise during this phase was with the Bowflex. I used 100 pounds of resistance and did 15 repetitions unless otherwise specified. My first set of exercises consisted of the bench press, chest fly, lying cable crossover, seated shoulder press, trunk rotation, and standing biceps curl (arm curl). My second set of exercises consisted of the seated triceps extension, seated resisted abdominal crunches at a resistance of 160, resisted reverse crunch at a resistance of 50, lying shoulder press, bilateral lat rows, back extensions with a resistance of 200, reverse curls at a resistance of 100, standing lateral shoulder raise at a resistance of 50, leg extensions at a resistance of 200, and leg curls at a resistance of 200. To end my second set, I would repeat the back extensions, seated resisted abdominal crunches at a resistance of 160, and resisted reverse crunch at a resistance of 50. I then repeated all of the exercises in my first set.

During the anabolic phase of my transformation, I mostly used the Bowflex to combine an aerobic workout with strength training. I dropped the three-mile run, push-ups, and sit-ups. I would use the same exercises as described in the catabolic phase. I increased the resistance on the seated triceps extension to 160, on the lying shoulder press to 160, on the seated resisted abdominal crunches to 250 (in order to build up the abs), bilateral rows to 160, and the second set of back extensions to 410. I would do the strength training part of my workout before and after the aerobic part. I did five exercises from my first set of the aerobic workout, except I would increase the tension and do ten repetitions. I would do two sets of these five exercises, one before the aerobic workout and one after. I started out at 300 on the bench press, chest fly, and shoulder press and then worked my way up to the maximum of 410. For the arm curls, I would start out at 200 and work my way up to 310. I increased the lying cable crossover to 200. Typically, toward the end of your strength training workout you will be able to do more than in the beginning. For example, there were days when I had difficulty pushing 350 on the Bowflex in the beginning of the workout. By the end of the workout I could push 360 or 370 with ease. This is because when you get the adrenaline pumping you are able to push more. It makes all the difference in the world. Typically, I am able to push more when the temperature is higher. I started out at my anabolic phase by bench pressing, chest flying, and shoulder pressing 310 pounds of resistance. My anabolic phase was finished when I reached 410 on all three of these exercises. I was arm curling 200 in the beginning of the anabolic phase and reached 310 toward the end of the anabolic phase. I used to be able to bench press, shoulder press, and chest fly 410 and arm curl 310, but after my catabolic phase, I couldn't. This is because I did not do strength training for about a year at 410 and 310, respectively. Much like any other muscle group in the body, the abs will hypertrophy when they are worked against tension. This will cause them to come out and become more prominent after your body fat percentage reaches low levels. The abs should be worked against high resistance in the anabolic phase.

I used protein shakes and supplements during both the catabolic and anabolic phases of my transformation. After my workout, I would have Amplified Wheybolic Extreme 60, a Megamen's Multivitamin, and a fish oil capsule. I would have L-carnitine before my workout.

Chapter 10

Which Supplements I Used and Which Supplements I Avoided

There are many supplements out there. Some of them are good, some are dangerous, and some do nothing.[21] One very important thing to note is that supplements are a *supplement* to proper dieting. They are not a *substitute* for proper dieting. If your diet is not good, then the supplement you take is most likely not going to do anything for you. Proper dieting will do more for you than any supplement will. I used certain supplements during my transformation. You should always check with your doctor before using supplements. Certain supplements may work for me, but they may not work for other people; their usefulness cannot really be determined by the experience of one person. Randomized clinical trials would have to be done.

One of the supplements I took was L-carnitine. I used to take 1500 mg a day. I would drink the protein shake Amplified Wheybolic Extreme 60, which has 500 mg of L-carnitine in it. It is the only protein shake I know of that has L-carnitine. It does contain an obesity additive called high fructose corn syrup, however; I didn't notice this until I was finished with my weight loss. Since I found it to be quite useful during

---

[21] Egras, A.M., Hamilton, W.R., Lenz, T.L., and Monaghan, M.S. (2011) An Evidence Based Review of Fat Modifying Supplemental Weight Loss Products. *Journal of Obesity.* Accessed from http://www.ncbi.nlm.nih.gov/pubmed/20847896

my weight loss, I continued to take it. I used L-carnitine throughout my weight loss and it worked well. L-carnitine brings stored fat into the mitochondria of cells and breaks it down into energy. I typically take it before my morning workout and take my protein shake after my workout. Some studies have shown that L-carnitine has no effect on weight loss.[22][23][24][25][26] One study found it to be useful.[27] So, it is even possible that I did not need the L-carnitine. My weight loss could have been attributed to diet and some aerobic exercise. After I finished my total body transformation, there were times where I could not exercise and had to diet to maintain the same weight and body fat percentage. I had to put myself back on L-carnitine to maintain my weight. So, again,

[22]   Villani, R.G., Gannon, J., Self, M., and Rich, P.A. L-Carnitine Supplementation Combined With Aerobic Training Does Not Promote Weight Loss in Moderately Obese Women. *International Journal Of Sports, Nutrition, Exercise, and Metabolism.* Accessed from http://www.ncbi.nlm.nih.gov/pubmed/10861338

[23]   Elmslie, J.L., Porter, R.J., Joyce, P.R., Hunt, P.J., and Mann, J.I. 2006 October) Carnitine Does Not Improve Weight Loss Outcomes in Valproate Treated Bipolar Patients Consuming and Energy Restricted Low Fat Diet. *Bipolar Disorder.* 8 (5 pt 1): 503-7. Accessed from http://www.ncbi.nlm.nih.gov/pubmed/17042889

[24]   Melton, S.A., Keenan, S.J., Stanciu, C.E., et al. (2005 March) L-Carnitine Supplementation Does Not Promote Weight Loss in Ovariectomized Rats Despite Endurance Exercise. International Journal for Vitamin and Nutrition Research. 75(2): 156-60. Accessed from http://www.ncbi.nlm.nih.gov/pubmed/15929637

[25]   Saldana Akoi, M., Rodriguez Amaral Almeida, A.L., Navarro, F., Bucido Pereira Costa-Rosa, L.F., and Pereira Bacurau, R.F. (2004) Carnitine Supplementation Fails to Maximize Fat Mass Loss Induced by Endurance Training in Rats. *Annals of Nutritional Metabolism.* 48(2):98-4. http://www.ncbi.nlm.nih.gov/pubmed/14988638

[26]   Brandsch, C. and Eder, (2002) Effects of L-Carnitine on Weight Loss and Body Composition of Rats Fed a Hypocaloric Diet. *Annals of Nutrition and Metabolism.* 46(5): 205-10. Accessed from K.http://www.ncbi.nlm.nih.gov/pubmed/12378044

[27]   Center, S.A., Warner, K.L., Randolph, J.F., Sunvold, G.D., and Vicker, J.R. (2012 July) Influence With Dietray Supplementation With L-Carnitine on Metabolic Rate, Fatty Acid Oxidation, Body Condition, and Weight Loss in Overweight Cats. American Journal of Veterinary Research. (2012 July) 73(7): 1002-15. Accessed from http://www.ncbi.nlm.nih.gov/pubmed/22738052

L-carnitine may have some usefulness when it comes to fat burning. I also took fish oil capsules and Mega Men's Multivitamin supplement. I always made sure my fish oil capsules were free of contaminants. For one month I also took the GNC supplement called Physioburn, which includes CLA. It did not really do anything for me. I lost ten pounds a month whether I used it or not. Studies show that CLA inhibits lipoprotein lipase, which prevents fat from being taken up and stored in fat cells.[28][29] Instead, it increases carnitine palmitoyl transferase enzyme, which takes fat into muscle cells to be broken down and used as energy. It should be taken with meals. It probably didn't do anything for me, because I didn't have a large amount of fat in my diet. CLA will probably not be useful for someone on a low fat diet. Some studies showed that

[28] Xu, X. Strokson, J., Kim, S., Sugimoto, K., Park, Y., and Pariza, M.W. (2003 March) Short-Term Intake of Conjugated Linoleic Acid Inhibits Lipoprotein Lipase and Glucose Metabolism But Does Not Enhance Lipolysis in Mouse Adipose Tissue. *Journal of Nutrition.* 133(3): 663-7. Accessed from http://www.ncbi.nlm.nih.gov/pubmed/12612134

[29] Zabala, A., Churruca, I., Fernandez-Quintela, A., Rodriguez, V.M., Macarulla, M., Martinez, J.A., and Portillo, M.P. (2006 June) Trans-10, Cis-12 Conjugated Linoleic Acid Inhibits Lipoprotein Lipase But Increases the Activity of Lipogenic Enzymes in Adipose Tissue from Hamsters Fed an Atherogenic Diet. British Journal of Nutrition. 95(6): 1112-9. Accessed from http://www.ncbi.nlm.nih.gov/pubmed/16768833

CLA is not useful in fat metabolism.[30][31][32][33][34][35][36][37] Other studies show that

30   Fedor, D.M, Adkins, Y., Newman, J.W., Mackey, B.E., and Kelley, D.S. (2013 February) The Effect of Doxosahexanonoic Acid on t10, t12-Conjugated Linoleic Acid Induced Changes in Fatty Acid Composition of Mouse, Liver, Adipose, and Muscle. *Metabolic Syndrome and Related Disorders.* 11(1):63-70. Accessed from http://www.ncbi.nlm.nih.gov/pubmed/23170930

31   Onokpoya, I.J., Posadski, P.P., Watson, L.K., Davies, L.A., and Ernst, E. (2012 March) The Efficacy of Long Term Conjugates Linoleic Acid (CLA) Supplementation on Body Composition In Overweight and Obese Individuals: A Systematic Review and Metanalysis of Randomized Clinical Trials. European Journal of Nutrition. (51(2): 127-34. Accessed from http://www.ncbi.nlm.nih.gov/pubmed/21990002

32   Plourde, M., Jew, S., Cunnane, S.C., and Jones, P.J. (2008 July) Conjugated Linoleic Acids: Whye the Discrepancy Between Animal and Human Studies? *Nutritional Review.* 66(7): 415-21. Accessed from http://www.ncbi.nlm.nih.gov/pubmed/18667017

33   Lasa, A., Churruca, I., Simon, E., Fernandez-Quintela, A., Rodriguez, V.M., and Portillo, M.P. (2008 December) Trans-10, Cis-12 Conjugated Linoleic Acid Does Not Increase Body Fat Loss Induced By Energy Restriction. *British Journal of Nutrition.* 100(6): 1245-50. Accessed from http://www.ncbi.nlm.nih.gov/pubmed/18507880

34   Larsen, T.M., Toubro, S., Gudmunsen, O., and Astrup, A. (2006 March) Conjugated Linoleic Supplementation Fro 1 Year Does Not Prevent Weight or Body Fat Regain. *American Journal of Clinical Nutrition.* 83(3):606-12. Accessed from http://www.ncbi.nlm.nih.gov/pubmed/16522907

35   Haugen, M. and Alexander, J. (2004 December 2) Can Linoleic Acids in Conjugated CLA Products Reduce Overweight Problems? *Tidsskr Nor Laegeforen.* 124(23): 3051-4. Accessed from http://www.ncbi.nlm.nih.gov/pubmed/15586185

36   Kamphuis, M.M., Lejeune, M.P., Saris, W.H., and Westerterp-Plantenga, M.S. (2003 October) Effects of Conjugated Linoleic Acid Supplementation After Weight Loss on Appetite and Food Intake in Overweight Subjects. *European Journal of Clinical Nutrition.* 57(10): 1268-74. Accessed from http://www.ncbi.nlm.nih.gov/pubmed/14506488

37   Kamphuis, M.M., Lejeune, M.P., Saris, W.H., and Westerterp-Plantenga, M.S. (2003 July) The Effects of Conjugated Linoleic Acid Supplementation After Weight Loss on Body Weight Regain, Body Composition, and Resting Metabolic Rates in Overweight Subjects. *International Journal Of Obesity and Related Metabolic Disorders.* 27(7): 840-7. Accessed from http://www.ncbi.nlm.nih.gov/pubmed/12821971

it is useful in fat metabolism.[383940414243]

Many supplements can be dangerous. The use of anabolic steroids is dangerous, as many health care providers will tell you. Anabolic steroids are an external source of testosterone. This source of testosterone shuts down the hypothalamic pituitary testicular axis. When this happens, the testes stop producing testosterone. Some of this excess testosterone turns into estrogen, causing male breast tissue to develop. I would not recommend the use of anabolic steroids. I would also not advise the use of creatine or products that contain creatine precursors. Renal damage has been associated with creatine. It should absolutely not be used in those with renal disease. In healthy people, creatine increases the size of muscles by forcing water into the muscles. This makes the muscles

---

[38] Chen, S.C., Lin, Y.H., Huang, H.P., Hsu, W.L., Houng, J.Y., and Hunag, C.K. (2012 May) Effect of Conjugated Linoleic Acid Supplementation on Weight Loss and Body Fat Composition in a Chinese Population. *Nutrition.* 28(5):559-65. Accessed from http://www.ncbi.nlm.nih.gov/pubmed/22261578

[39] Norris, L.E., Collene, A.L., Asp, M.L., Hsu, J.C., Liu, L.F., Richardson, J.R., Li, D., Bell, D., Osei, K., Jackson, R.D., and Belury, M.A. (2009 September) Comparison of Dietary Conjugated Linoleic Acid with Safflower Oil on Body Composition in Obese Postmenopausal Women With Type 2 Diabetes Mellitus. *American Journal of Clinical Nutrition.* 930(4)-68-76. Accessed from http://www.ncbi.nlm.nih.gov/pubmed/19535429

[40] Park, Y., Albright, K.J., Storkson, J.M., Liu, W., and Pariza, M.W. (2007 October) Conjugated Linoleic Acid Prevents Body Fat Accumulation and Weight Gain in an Animal Model. *Journal of Food Science.* 72(8): S612-7. Accessed from http://www.ncbi.nlm.nih.gov/pubmed/17995628

[41] Whigham, L.D., Watras, A.C., and Schoeller, D.A. (2007 May) Efficacy of Conjugated Linoleic Acid for Reducing Fat Mass: A Meta-Analysis in Humans. *American Journal of Clinical Nutrition.* 85(5): 1203-11. Accessed from http://www.ncbi.nlm.nih.gov/pubmed/17490954

[42] Gaullier, J.M., Halse, J., Hoivik, H.O., et al. (2007 March) Six Months Supplementation With Conjugated Linoleic Acid Induces Regional-Specific Fat Mass Decreases in Overweight and Obese. *British Journal of Nutrition.* 97(3):550-60. Accessed from http://www.ncbi.nlm.nih.gov/pubmed/17313718

[43] Gaullier, J.M., Halse, J., Hoyue, K., Kristiansen, K., Fagertun, H., Vik, H., and Gudmunsen, O. (2005 April) Supplementation with Conjugated Linoleic Acid from 24 Months is Well Tolerated By and Reduces Body Fat Mass in Healthy, Overweight Humans. *Journal of Nutrition.* 135(4): 778-84. Accessed from http://www.ncbi.nlm.nih.gov/pubmed/15795434

more prone to strains, sprains, tears, and other types of injuries. A lot of weight loss supplements contain stimulants. I do not tolerate stimulants very well, and they can cause hypertension. Weight can be lost in a healthy manner without these stimulant weight loss pills. I would not advise the use of stimulants for weight loss, because they are really not needed.

There are many protein bars out there, but most of them are nothing more than a regular candy bar with protein in it. They contain lots of sugars and fat. The nutritional value of a protein bar can be determined by its nutritional label. There are very few protein bars that are actually good for you. Two protein bars that do have a good nutritional label are Quest Bars and Chocolite Protein Bars. These bars have a lot of protein and fiber and are low in fat and sugar. MHP Protein Pak Power Pudding has a good nutritional label as well as a good taste.

Protein shakes are also like protein bars. You need to choose one with low fat and low sugar. Many protein shakes aren't good because they have too much fat and sugar. GNC's Amplified Wheybolic Extreme 60 is the protein shake I use. It is low fat and low sugar and is the only protein shake I know of that has L-carnitine. After my catabolic phase, I noticed it had some high fructose corn syrup in it, an obesity additive. I used this shake throughout my total body transformation and lost weight with it, despite the fact that it has an obesity additive in it. Another protein shake I use is Vitamin World's Whey Protein Isolate. It is a protein shake that has a good nutritional label as well. However, it does not contain L-carnitine. I also used Jay Robb Whey Protein. Protein shakes should be taken after a workout, as protein can be used to help build and repair muscles.

I also used Force Factor during my anabolic phase. Force Factor may help some people build muscle. Again I am not sure how helpful this was for me in building up muscles. I did do eight years of strength training before my total body transformation. I started out bench pressing, shoulder pressing, and chest flying 310. I am not sure how much muscle gain during the anabolic phase was attributed to Force Factor. I could not find any articles claiming that nitric oxide supplements increase muscle mass.

I also used Formedrol Extreme, an aromatase inhibitor that decreases the body's estrogen levels. It prevents the body from converting estrogen to testosterone, so the testosterone/estrogen ration is higher. I wanted to use this to boost testosterone levels. However the testosterone levels were already boosted with the fat loss. Since I purchased it, I decided to use it in the anabolic phase. This is not for everyone. You should always consult with a physician before using this. It can cause an unfavorable LDL and HDL profile. It should never be used for more than eight weeks. I only used it for six weeks and I'm not sure how useful it was. Before I ever even used Force Factor or Formedrol Extreme, I was bench pressing, chest flying, and shoulder pressing 410 and arm curling 310. I was doing this for a few years before my total body transformation, so it is more likely that the muscle mass I gained during my anabolic phase was due to my natural abilities and the fat loss causing an increase in testosterone levels. The muscle gain was not likely due to these supplements.

When I completed my total body transformation, I eventually weaned myself off most of the supplements. I continued to take fish oil capsules, multivitamins, protein bars, protein shakes, and protein pudding. I relied mostly on proper dieting and exercise to maintain the results of my transformation. I took some zinc supplements, raspberry ketone supplements, and indole 3 carbinole supplements after my transformation, but I eventually weaned myself off them as well. Two studies showed that raspberry ketones were useful in fat burning.[44][45]

Walden Farm's creates a lot of no-calorie foods including ketchup, barbeque sauce, salad dressing, chocolate/caramel dips and syrups, and so forth. I started substituting Walden Farm's products for brand that I used to buy. All of their products can be found on their website at waldenfarms.com. I do not like the taste of some Walden Farms products. An alternative to Walden Farms would be Greenway Organic Products. Although they are not calorie free, they do have good nutrition labels compared to other products.

---

[44] Park, K.S. (2010 October) Raspberry Ketone Increases Both Lipolysis and Fatty Acid Oxidation in 3T3-L1 Adipocytes. *Planta Medica.* 76(15): 1654-8. Accessed from http://www.ncbi.nlm.nih.gov/pubmed/20425690

[45] Park, K.S. (2010 October) Raspberry Ketone Increases Both Lipolysis and Fatty Acid Oxidation in 3T3-L1 Adipocytes. Planta Medica. 76(15): 1654-8. Accessed from http://www.ncbi.nlm.nih.gov/pubmed/20425690

Chapter 11

Foods That Are Known to Burn Fat

W hen you search on the Internet, many foods come up that are believed to burn fat. Again, before you start eating these fat-burning foods, you should always check with your doctor because they may not be good for you. For example, a person with hypertension should not be eating anything with stimulants in it. Let's take a look at some of these foods. One that comes up all the time is dark chocolate. Dark chocolate is said to burn fat because the cocoa beans in dark chocolate can increase your metabolism; they contain a stimulant called *theobromine*. However, I have never seen a dark chocolate bar with low fat and sugar. Typically, dark chocolate bars that are sugar free will have lots of fat in them. Those that are low in fat will have a lot sugar in them. So I stay away from dark chocolate.

Green tea is also well known to burn fat. I used green tea in the later stages of my weight loss. Many people who cannot tolerate caffeine like myself can have decaffeinated green tea. Make sure the green tea is CO_2-decaffeinated because this preserves the green tea nutrients and does not cause contamination with ethyl acetate. Salada Green Tea is one brand of tea that is CO2 decaffeinated.

Different nuts are also known to burn fat. Nuts have a lot of monounsaturated fats, polyunsaturated fats, and proteins. However nuts also have saturated fats. The way nuts work is that they cause a feeling of fullness, which makes people not want to eat more. I occasionally ate nuts during my weight loss. Nuts should be eaten in moderation as they have saturated fats and are also high in calories. When losing weight, you need to eat low-calorie foods.

Different types of berries are known to be good for losing fat. I ate berries a lot during my catabolic phase, as I mentioned before. I also had decaffeinated green tea with the berries, a combination that helped me lose a lot of weight. Berries by weight are mostly water, which just gets urinated out. There is also a lot of fiber in berries, which mostly just passes through the gut. This fiber also decreases the glycemic index of the berries. Then there is the digestible part of the berry, which requires a lot of energy for the body to burn.

Oatmeal is another food that is advertised as a fat burning food. You must always read the label to see how much sugar it has. Many oatmeal brands have a lot of sugar and may also be low in fiber. You must always read the label when purchasing oatmeal. If you have oatmeal, it's a good idea to sprinkle cinnamon on it, not sugar. Cinnamon is also known to be a good fat burning agent.

Avocados are another food that can help you lose fat. Avocados are high in fiber and monounsaturated fat. I ate some avocados in the beginning of my weight loss, but did not continue to eat them because I did not like the taste. Many sources say they are good for fat loss.

Different types of peppers are said to cause people to lose weight. They speed up your metabolism. This is why when I make food on the grill, I always add spices to it. I happen to like spicy foods. It also adds more flavor to the meat when I eat it.

Certain meats such as turkey, salmon, and chicken breast are also known to burn fat. However, in the catabolic phase I would not recommend eating more than one-half pound of meat with each meal. More fruits and vegetables should be eaten with meats during the catabolic phase. The fiber from the fruits and vegetables will prevent the take-up of fat when the meat is digested.

Many people say that eggs are good for fat loss. There is some controversy as to whether or not the whole egg should be eaten or just the egg white. Some people argue that the egg yolk also has protein, and much of the fats in eggs yolks are unsaturated fats. They also say that eggs contain CLA. Some say that eggs increase cholesterol. When a food is considered to be good for fat loss by some health experts and bad by others, I would just avoid it. Besides, I have never been a fan of eggs anyway.

They say olive and canola oils are the best oils when it comes to burning fat. It would make sense because of the high amount of unsaturated fats in them. I always use these oils for cooking. Make sure the oil you use has no trans fats; many vegetable oils have them.

Peanut butter is often advertised as a fat burning food. This makes sense, considering peanut butter is a product of peanuts and has a lot of unsaturated fats. Be careful, though, because it also contains saturated fat. It should be eaten in moderation.

Spinach and many other green vegetables are also good for fat burning. This is because they are low calorie foods. I ate spinach a lot during the course of my weight loss. I would add a lot of spices and also mushrooms to the spinach.

Beans, chickpeas, and lentils are also said to be good for fat burning. They make a good substitute for meats because they have more fiber and less saturated fat. They are also a good source of protein. I ate them a lot during my catabolic phase and found them to be tasty.

Low fat dairy foods are also advertised as being good for fat burning. I have seen fat free and sugar free cheese, but I have never seen fat free and sugar free yogurt and milk. For this reason, I avoided milk and yogurt in my fat loss diet. They both have a low glycemic index, which would make them okay for fat loss (assuming they are fat free). Also, the calcium contained in yogurt and milk makes them okay for fat loss. My diet was stricter than usual. Most of the sugar I consumed came from fruits and vegetables, but the ones I ate always had a low glycemic index.

Broccoli also shows up on the list of fat burning food. Broccoli is low in calories, contains a lot of fiber, and also has indole 3 carbinole, which increases testosterone. Again, most fruits and vegetables with a low glycemic index are low in calories and good for fat burning. They are the core of the diet in the catabolic phase.

The African mango has also been said to cause weight loss. I am not sure where to purchase African mangos here in America. However, they sell an African mango extract at health food stores, but is has received mixed reviews. One study found it to be effective, but another did not.[46][47] It causes weight loss in some and not in others. I took this supplement in the last month of my catabolic phase and for a short time after my total body transformation, but it did not really help me lose weight.

[46] Ross, S.M. (2011 July-August) African Mango (IGOB131): A Proprietary Seed Extract of Irvingia Gabonesis is Found to be Effective in Reducing Body Weight and Improving Metabolic Parameters in Overweight Humans. *Holistic Nurse Practitioner.* 25(4): 215-7. Accessed from http://www.ncbi.nlm.nih.gov/pubmed/21697664

[47] Onakpoya, I, Davies, J.L., Posadski, P., and Ernst, E. (2013 March) The Efficacy or Irvingia Gabonensis Supplementation in the Management of Overweight and Obesity: A Systematic Review of Randomized Controlled Trails. *Journal of Dietary Supplements.* 10(1): 29-38. Accessed from http://www.ncbi.nlm.nih.gov/pubmed/23419021

Chapter 12

A Closer Look at Body Fat Percentage

We talked a little bit about body fat percentage in chapter 5. In my opinion, this is the most critical measurement. The important thing to know here is that even if you gain weight, your body fat percentage can remain the same. Your body fat percentage typically decreases when abdominal measurements (abdominal circumference in men and narrowest point/widest point circumference in women) decrease. When the neck circumference increases, the body fat percentage typically decreases. This can be wrong, however, because fat can also be gained around the neck. The most critical measurement is the abdominal circumference. If you have gained a few pounds but your abdominal circumference is still the same, then it is not likely that the pounds you gained are attributed to fat gain. It could be due to muscle gain, water weight, food left over in the gut, or other reasons. When you are finished with your catabolic phase and are in your anabolic phase, you will most likely note that your weight is increasing. This is going to be due to muscle. To make sure you are not gaining fat, you would have to maintain your abdominal measurements. Fat has a tendency to store itself first in the abdomen in men and in the hips and the buttocks in women. For this reason these abdominal measurements are very critical in determining body fat. If you are in your catabolic phase and you work out your abs and your obliques, this may cause your abdominal measurements to increase. However, this will be a gain in muscle—not a gain in fat. If you note that your abs and obliques are getting more pronounced, then it is likely due to muscle gain and not fat gain, even though your abdominal measurements are increasing.

Another thing to note is that body fat percentage only takes into consideration the amount of body fat, not muscle mass. So a slim person may weigh a lot less than a muscular person and may have the same body fat percentage. Women should never go below 10 percent body fat, because 10–13 percent is considered essential fat. Anorexics typically have body fat levels less than this, and that can lead to infertility and hormonal problems. Essential fat for men is around 2–5 percent. Typically, only bodybuilders reach these levels. No one should ever go below his or her essential fat range.

Some websites do a very good job of describing body fat percentage with pictures.[484950] The websites referenced below show pictures of people at various body fat percentages. One website describes classifications of body fat percentages in men.[51] They start with *Full House*, which is over or near 20 percent. At this stage, there is little muscle definition and not much separation between muscle groups. The *Hard* stage is at around 15 percent. The abs are not visible, but some muscle definition appears in the upper arms. The *Cut* stage is at around 12 percent. This is where the chest and back has muscle separation. The abs can be seen faintly in this stage. At 10 percent there is the *Defined* stage. The arms, chest, back, and legs appear. The abs can be seen during flexion. The *Ripped* stage is at 7–9 percent. In this stage there is chest and back separation, arm vascularity, and very prominent abs. At 4–7 percent body fat there is the *Shredded* stage in which muscle striations occur on flexion. Vascularity appears on the lower abdomen and legs. Then there

---

[48] Leigh Peele (2010 Februray 21) Body Fat Pictures and Percentages. Accessed from http://www.leighpeele.com/body-fat-pictures-and-percentages

[49] This is Why Your Jacked (n.d.)http://thisiswhyyourejacked.com/a-body-fat-percent-picture-guide/

[50] Clark, S. (2010 September 29) Many People Do Not Have Access to Fancy Equipment to Determine Their Body Fat. Here are Some Categories to Help Explain Fat Percentage. Accessed from http://www.bodybuilding.com/fun/body_fat_categories.htm

[51] This Why Your Jacked (n.d.) http://thisiswhyyourejacked.com/a-body-fat-percent-picture-guide/

is the final stage, *Sliced*, at 3 percent body fat. Striations and vascularity appear everywhere. This last stage is not a very healthy stage because it has low subcutaneous water levels. This condition can only be sustained for a few hours.

Chapter 13

Some Recipes for Healthy and Tasty Meals

pretty much changed my whole diet with my total body transformation. I learned to cook some new things as well as buy some new dishes from stores. The foods I ate were healthy as well as tasty. We will talk about how to make some of these dishes. The dishes in the catabolic and anabolic phases of the transformation were similar, but in the catabolic phase there was a smaller portion of protein and more fruits and vegetables. I would not eat more than one-half pound of meat with each meal. My portions of lentils, beans, and chickpeas were smaller. More low calories foods were eaten during the catabolic phase. In the anabolic phase I ate more protein; I ate fruits and vegetables as well, but the majority of my diet was protein.

I liked to make breaded salmon, chicken breast cutlets, or cod fish. I would dip these meats into Kraft Tuscan House Italian Dressing and Marinade then I would cover with Shake and Bake Bread Crumbs. I would spray the pan with Omega-3 Pam and then I would cook it in the oven for forty-five minutes. This would really taste good, especially with hot sauce. At times I would make chicken parmigiana cutlets using this recipe. I would make the chicken cutlets and then cover them with either Walden Farms Tomato and Basil Sauce or Greenway Organic Roasted Garlic Tomato Sauce. I would melt Kraft's Shredded Fat Free Mozzarella Cheese on it.

Another recipe I made was on the grill. I would make up my own spice mix with salt, garlic powder, onion powder, paprika, crushed dried peppers, and garam masala (from an Indian food store). I would add two spoonsful of this to Perdue's Ground Chicken. I would mix and make chicken burgers. I would always squeeze out as much fat as I could when grilling the burgers. I would make my own marinade, which consisted of any of the Walden Farm's Barbeque Sauces, lemon juice, and the above spice mix. I would marinade drumsticks, chicken kabobs, or salmon kabobs and then cook them on the grill. I would always take the fat off the drumsticks and chicken kabobs.

I would also make my own taco platter. I would not eat taco shells. The mix would be made without taco shells. I would cut up tomatoes and onions and then add Old El Paso Taco Seasoning with Perdue's Ground Chicken. I would always drain the fat out of this. I would add Kraft's Shredded Fat Free Cheddar Cheese and Old El Paso's Taco Sauce and then mix all of this together to get a good tasting taco platter.

There were some foods that I would eat a lot of, but could not cook myself. These were mainly Indian dishes. I would either buy them from the Indian store or have my aunt cook them for me. They were spicy black lentils (*kali dhal*), kidney beans (*rajma*), green peas (*matar* without the cheese or *paneer*), chickpeas (*chole masala*), and spinach (*saag* without the cheese or paneer). I would also buy dried chickpeas with spices and a dish called *papad*. They are both high in protein. Dried chickpeas work well as an appetite suppressant. Papads are mashed up lentils that are cooked to form a crunchy shell. They have three grams of protein and two grams of fiber per serving. They have low sugar and fat.

Chapter 14

How to Maintain the Results of the Total Body Transformation

After you finish the catabolic and anabolic phases, you get to the third phase of the total body transformation: the maintenance phase. Maintaining the results of the total body transformation is a lifetime commitment. The most important thing to do to maintain your results is to eat right. Avoid saturated fats, trans fats, sugars, alcohols, and foods with a high glycemic index. Proteins from lean sources, fiber, monounsaturated fats, polyunsaturated fats, and food with a low glycemic index should be consumed. I did not eat medium glycemic index sugars during my total body transformation, but I ate them afterward and did not gain weight with them. Try to eat foods with a high nutritional value. On days when you eat more protein, your weight will typically go up. On days when you eat more fruits and vegetables, your weight will come down. This is the way I keep my weight consistent.

As far as exercise is concerned, it all depends on what you are trying to do. If you are trying to stay slim, then you need more of an aerobic workout. If you want to maintain muscle mass and muscle tone, then add strength training to your aerobic workout. My workout was a Bowflex workout. I would do a forty-five-minute workout that combined aerobic exercise with strength training. I would start out by doing strength training followed by an aerobic workout and then end it with some strength training. Workouts can be done two to four times a week. Again, the most important component is diet.

You should monitor your weight and body fat percentage. The most critical component of the body fat percentage is the abdominal measurements. If fat starts to accumulate on the body, it will accumulate there first and will be detected when you measure your body fat percentage. Since fat has the tendency to accumulate on the abdomen in men, it would make sense that the abdominal circumference at the level of the umbilicus, which is used to measure body fat percentage, would be the measurement to keep an eye on in men. In women, fat starts to accumulate more on the hips and buttocks, so the narrowest abdominal circumference and widest hip circumference, which are used to measure body fat percentage, would be the measurements to keep an eye on in women. My diet after my transformation consisted of a protein source combined with fruits and vegetables with a low or medium glycemic index. If I found I was gaining a little weight, I would eat more of the low glycemic index fruits and vegetables and less of the protein source. This would decrease my weight by one or two pounds. I monitored my weight and body fat measurements, and if I could not exercise because my Bowflex was broken, then I would have to rely on my diet to keep the weight down. I would even have to go back on L-carnitine sometimes. It may not work for everyone, but it worked for me. I did gain a few more pounds of muscle in the long run. I kept my weight in the mid to low 130s. My abdominal circumference, at the level of the umbilicus, dropped to as low as twenty-five inches at times. The loose skin in my lower abdomen became tighter as time went by. After my transformation, I did the same exercise routine I did during the anabolic phase, which combined strength training with an aerobic workout.

The common mistake people make is that they lose the weight and then go back to their unhealthy lifestyle—and they end up gaining the weight back again. In order to keep the weight off you must maintain a healthy diet.

You have seen the *before* picture in the beginning of this book. I did not have too many beach pictures before my transformation. I will show you pictures after the catabolic phase and then after the anabolic phase. Bear in mind these are not professionally taken pictures manipulated by the camera, and these are not modeling pictures. They are honest, straightforward pictures. Notice the increased muscle tone and definition. The evidence that this weight loss took place quickly is the loose skin. Anytime you lose a lot of weight in a short period of time you will have loose skin. It can take years for the skin to adjust to the new body shape. Sometimes creams or lotions with collagen and elastin help. Skin is very compliant. We see this compliancy in pregnancy. When a woman gets pregnant the skin expands to accommodate the gravid uterus. When the woman delivers, the skin contracts back down to accommodate the postpartum shape of the body. This can take some time.

Sometimes surgery may be required to fix loose skin. It may be needed if a lot of weight is lost, if it is lost over a short person of time, or if it is an elderly person. Bear in mind that I went from a BMI and body fat percentage that was considered obese down to an essential body fat percentage and normal BMI. When you look at the profile views you will see just how much abdominal fat came off.

7½ months later

Post catabolic phase

The subject (me)

Weight: 119 lbs. (81 lbs. weight loss)

Abdominal circumference at the level of the umbilicus: 27″

Waist: 28″

Body fat percentage: 4.44%

BMI: 18.4

No more obesity, hypertension, hyperlipidemia, snoring, or backache

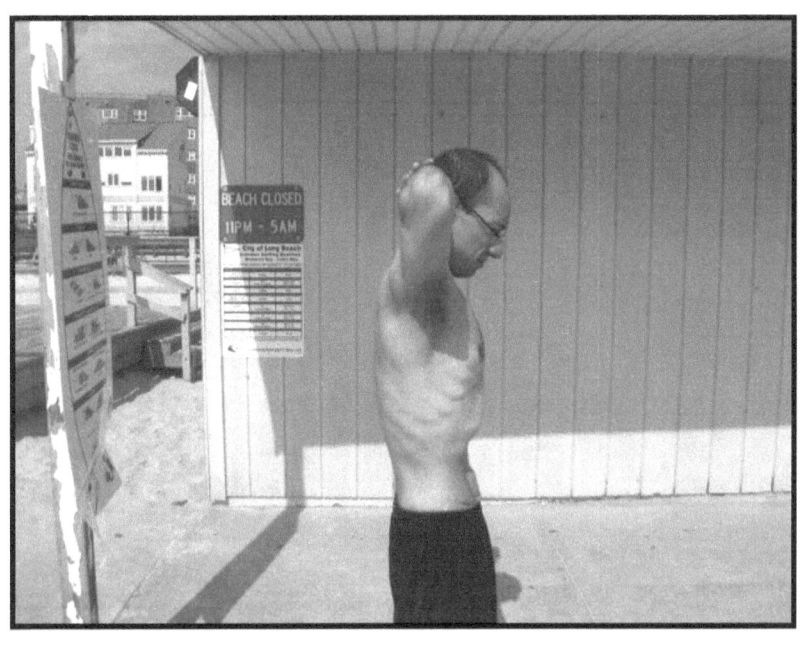

Two weeks later

Post anabolic phase/total body transformation complete

The subject (me)

Weight: 132 lbs. (13 lbs. muscle gain)

Abdominal circumference at the level of the umbilicus: 27″

Waist: 28″

Body fat percentage remains at 4.44%

BMI: 20.4

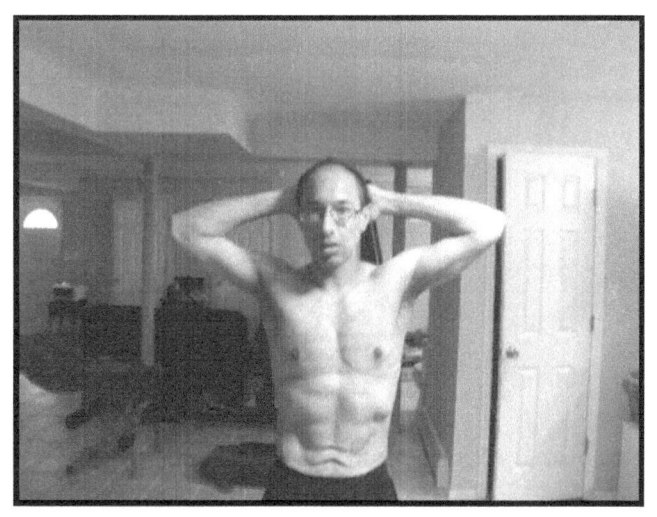

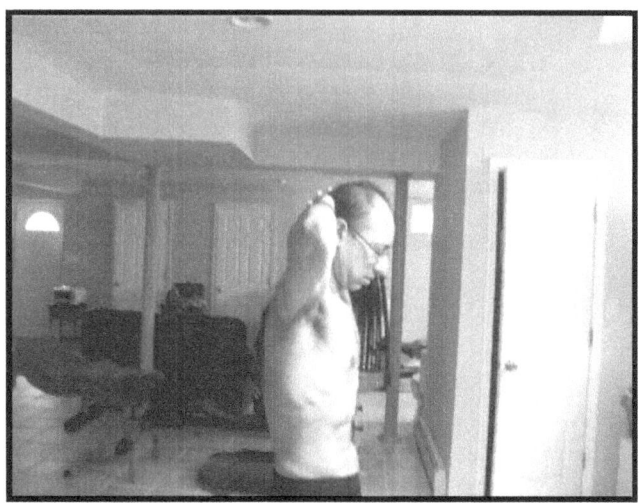

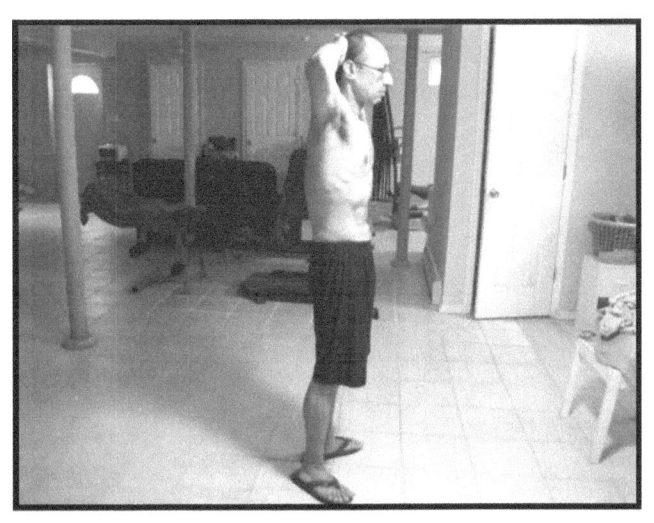

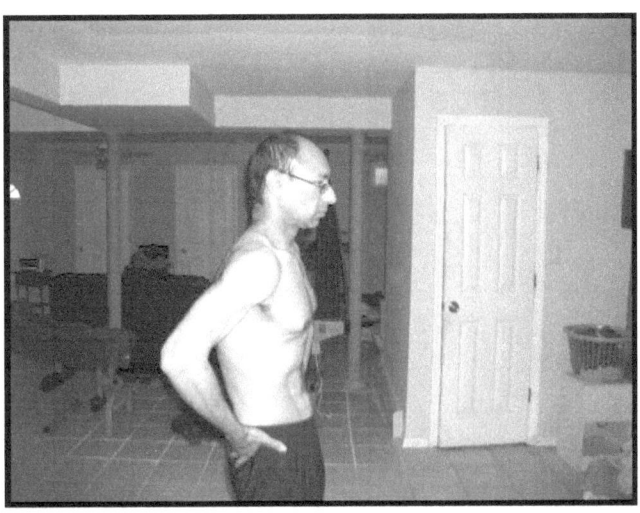

A few months later the loose skins shows improvement, and it continues to get better over time.

Always remember, in order to get extreme results, you have to set extreme goals!

www.ingramcontent.com/pod-product-compliance
Lightning Source LLC
Chambersburg PA
CBHW022123170526
45157CB00004B/1734